More praise for *Intelligence and ....*

"If intelligence were deeply encoded in our genes, that would lead to the depressing conclusion that neither schooling nor antipoverty programs can accomplish much. Yet while this view of I.Q. as overwhelmingly inherited has been widely held, the evidence is growing that it is, at a practical level, profoundly wrong. Richard Nisbett, a professor of psychology at the University of Michigan, has just demolished this view in a superb new book, *Intelligence and How to Get It*, which also offers terrific advice for addressing poverty and inequality in America."
—Nicholas D. Kristof, *New York Times*

"A meticulous and eye-opening critique of hereditarianism . . . its real value lies in Nisbett's forceful marshaling of the evidence, much of it recent, favoring what he calls 'the new environmentalism,' which stresses the importance of nonhereditary factors in determining I.Q." —Jim Holt, *New York Times Book Review*

"Impressively accessible and engaging."
—Steve Ruskin, *Rocky Mountain News*

"Nisbett shows why the key claims of the hereditarian camp are wrong—why intelligence isn't fixed at birth and why racial differences in IQ and educational achievement aren't rooted in genetics. . . . Marshals supporting evidence for his case from all corners of the scientific universe. . . . If we want to close those [achievement] gaps, Nisbett's book shows us what we can do."
—David L. Kirp, *American Prospect*

"Whether intelligence is largely determined by genetics or environment has long been hotly contested. Nisbett, a University of Michigan psychology professor, weighs in forcefully and articulately, claiming that environmental conditions almost completely overwhelm the impact of genes. He comes to this conclusion through a careful statistical analysis of a large number of studies and also demonstrates how environment can influence not only IQ measures but actual achievement of both students and adults. . . . Nisbett builds a very strong case that measured IQ differences across racial, cultural and socioeconomic boundaries can easily be explained without resorting to hereditary factors."

—*Publishers Weekly*

"Nisbett's book is a hugely important analysis of the determinants of IQ, and it deserves to weigh heavily on educational policy. The breadth and lucidity of the discussion make the work a 'must-read' for social scientists and the general public."

—Daniel Osherson, professor of psychology,
Princeton University

"This book is an excellent antidote to the prevailing view that our genes forbid enhancing intelligence and learning. Nisbett's lucid mind and style make it accessible to a wide audience. Its particular strength is its insistence that cases in which cognitive abilities have been enhanced take precedence over mathematical demonstrations that this is impossible."

—James R. Flynn, author of *What Is Intelligence?*

Intelligence

& HOW TO GET IT

Intelligence
AND HOW TO GET IT

Why Schools and Cultures Count

RICHARD E. NISBETT

W. W. NORTON & COMPANY
New York • London

For information about permission to reproduce selections from this book,
write to Permissions, W. W. Norton & Company, Inc., 500 Fifth Avenue,
New York, NY 10110

For information about special discounts for bulk purchases, please contact
W. W. Norton Special Sales at specialsales@wwnorton.com or 800-233-4830

Manufacturing by Courier Westford
Book design by Charlotte Staub
Production manager: Anna Oler

Library of Congress Cataloging-in-Publication Data

Nisbett, Richard E.
Intelligence and how to get it : why schools and cultures count /
Richard E. Nisbett. — 1st ed.
p. cm.
Includes bibliographical references and index.
ISBN 978-0-393-06505-3 (hardcover)
1. Intellect 2. Intelligence 3. Schools. 4. Culture. I. Title.
BF431.N57 2009
153.9—dc22   2008044477

ISBN 978-0-393-33769-3 pbk.

W. W. Norton & Company, Inc.
500 Fifth Avenue, New York, N.Y. 10110
www.wwnorton.com

W. W. Norton & Company Ltd.
Castle House, 75/76 Wells Street, London W1T 3QT

1 2 3 4 5 6 7 8 9 0

For

LEE ROSS

CONTENTS

ACKNOWLEDGMENTS

THE WRITING OF THIS BOOK, and some of the research reported in it, was supported by the National Science Foundation under Grant No. 0717982 and by the National Institute on Aging under Grant No. 1RO1AG029509-01A2. Neither agency should be assumed to endorse the views represented in the book. The psychology department of Columbia University and the Russell Sage Foundation contributed valuable resources and facilities.

Many people generously provided ideas and criticism that improved this book—though none of them is responsible for any errors in it. These people include Joshua Aronson, Douglas Besharov, Clancy Blair, Jeanne Brooks-Gunn, Hannah Chua, William Dickens, James R. Flynn, Phillip Goff, Richard Gonzalez, David Grissmer, Diane Halpern, Lawrence Hirschfeld, Earl Hunt, Shinobu Kitayama, Matt McGue, Walter Mischel, Randolph Nesse, Dan Osherson, Daphna Oyserman, Denise Park, Richard Rothstein, Peter Salovey, Kenneth Savitsky, Edward E. Smith, Jacqui Smith, Claude Steele, Robert Sternberg, Eric Turkheimer, Barbara Tversky, Jane Waldfogel, and Oscar Ybarra. I am grateful to my agents, John Brockman and Katinka Matson, for representing me and for facilitating the public's access to scientific writing. I thank my editors—Angela von der Lippe, Erica Stern, and Mary Babcock—for excellent work in bringing the manuscript to publication. Laura Reynolds provided assistance in preparing the manuscript. Katherine Rice provided invaluable help in the form of library research and vigorous and constructive criticism. Susan Nisbett made excellent suggestions and provided sage advice.

Lee Ross contributed to this book, as he has to all my other projects since I first met him in graduate school. For his intellectual stimulation and his friendship, this book is dedicated to him.

Intelligence

&

HOW TO GET IT

CHAPTER ONE

Varieties of Intelligence

> By intelligence the psychologist [means] inborn, all-around
> intellectual ability . . . inherited, not due to teaching or train-
> ing . . . uninfluenced by industry or zeal.
> —Sir Cyril Burt and colleagues (1934)

I BEGAN HAVING TROUBLE with arithmetic in the fifth grade, after I missed school for a week just when my class took up fractions. For the rest of elementary school I never quite recovered from that setback. My parents were sympathetic, telling me that people in our family had never been very good at math. They viewed math skills as something that you either had or not, for reasons having mostly to do with heredity.

My parents probably were not aware of the psychological literature on the question of intelligence, but they were in tune with it. Many if not most experts on intelligence in the late twentieth-century believed that intelligence and academic talent are substantially under genetic control—they are wired in and more or less unfold in any reasonably normal environment. Such experts were suspicious about the likely success of any effort to improve intelligence and were not surprised when interventions such as early childhood education failed to have a lasting effect. They were quite dubious that people could become smarter as the result of improvements in education or of changes in society.

But the results of recent research in psychology, genetics, and neuroscience, along with current studies on the effectiveness of educational interventions, have overturned the strong hereditarian position on intelligence. It is now clear that intelligence is highly modifiable by the environment. Without formal education a person is simply not going to be very bright—whether we measure intelligence by IQ tests or any other metric. And whether a particular person's IQ—and academic achievement and occupational success—is going to be high or low very much depends on environmental factors that have nothing to do with genes.

There are three important principles of this new environmentalism:

1. Interventions of the right kind, including in schools, can make people smarter. And certainly schools can be made much better than they are now.
2. Society is making ever greater demands on intelligence, and cultural and educational environments have been changing in such a way as to make the population as a whole smarter—and smart in different ways than in the past.
3. It is possible to reduce the IQ and academic achievement gap that separates people of lower economic status from those of higher economic status, as well as the gap between the white population and some minority groups.

The basic message of this book is a simple one about the power of the environment to influence intelligence potential, and more specifically about the role that schools and cultures play in affecting the environment. The accumulated evidence of research, much of it quite recent, provides good reason for being far more optimistic about the possibilities of actually improving the intelligence of individuals, groups, and society as a whole, than was thought by most experts even a few years ago.

On the other hand, just as there are laypeople and experts who are wrongly convinced that intelligence is mostly a matter of genes, there are laypeople and experts who have mistaken and sometimes overly optimistic ideas about the sorts of things that can improve intelligence and academic performance. One of the goals of this book is to present evidence on which interventions are most effective.

The chapters that follow emphasize that societal and cultural differences among groups have a big impact on intelligence and academic achievement. People of lower socioeconomic status have lower average IQs and achievement for reasons that are partly environmental—and some of the environmental factors are cultural in nature. Blacks and other ethnic groups have lower IQs and achievement for reasons that are entirely environmental. Most of the environmental factors relate to historical disadvantages but some have to do with social practices that can be changed.

Culture can also confer advantages for the development of intelligence and academic achievement. Some cultural groups have distinct intellectual advantages, on average, over the mainstream white population. These include people with East Asian origins and Ashkenazi Jews. Later, I discuss what these advantages are due to and whether some of them might be adopted by others who would like to increase their own intelligence and academic achievement.

Finally, I present ways of improving intelligence as suggested by new scientific findings.

Nearly everything in this book is readily understood without any particular technical knowledge. But it might be helpful to be familiar with statistics, so I provide an appendix defining some terms. Readers who want to brush up on their statistics might also want to look at the appendix. The concepts discussed there are normal distribution, standard deviation, statistical significance, effect size (in standard deviation terms), correlation coefficient, self-selection, and multiple-regression analysis.

Note that I have a somewhat atypical aversion to multiple-regression analysis, in which a number of variables are measured and their association with some dependent variable is examined. Such analyses can give a false impression of the degree to which causality can be inferred, and I refer to them only rarely and always with skepticism. Those of you who would like to see the basis of my prejudice can look at the part of Appendix A that deals with it.

To get us started, in this chapter I define intelligence, discuss how it is measured, present evidence on the two different kinds of analytic intelligence assessed by IQ tests, and discuss the types of intelligence not measured by IQ tests. I also examine how well IQ predicts academic achievement and occupational success, the types of intelligence not tapped by IQ, and important aspects of motivation and character.

Defining and Measuring Intelligence

A definition of intelligence by Linda Gottfredson is a good place to start:

> [Intelligence is] a very general mental capability that, among other things, involves the ability to reason, plan, solve problems, think abstractly, comprehend complex ideas, learn quickly and learn from experience. It is not merely book learning, a narrow academic skill, or test-taking smarts. Rather it reflects a broader and deeper capability for comprehending our surroundings—"catching on," "making sense" of things, or "figuring out" what to do.

Experts in the field of intelligence agree virtually unanimously that intelligence includes abstract reasoning, problem-solving ability, and capacity to acquire knowledge. A substantial majority of experts also believe that memory and mental speed are part of intelligence, and a bare majority include in their definition general knowledge and creativity as well.

These definitions leave out some aspects of intelligence that people in other cultures would be likely to include. Developmental psychologist Robert Sternberg has studied what laypeople in a large number of cultures think should be counted as intelligence. He finds that a good many people include social characteristics, such as ability to understand and empathize with other people, as aspects of intelligence. This is especially true of East Asian and African cultures. In addition, East Asian understanding of intelligence emphasizes the pragmatic, utilitarian aspects more than Western views do, which are more likely to value the search for knowledge for its own sake, whether or not it has any obvious immediate uses.

Intelligence is often measured by IQ tests. The *Q*, by the way, stands for *quotient*. The original IQ tests, which were devised for schoolchildren, defined intelligence as mental age divided by chronological age. By that definition, a ten-year-old with a test performance characteristic of twelve-year-olds would have an IQ of 120; one with a mental age typical of eight-year-olds would have an IQ of 80. But modern IQ tests arbitrarily define the mean of the population of a given age as being 100 and force the distribution around that mean to have a particular standard deviation—usually 15. So a person whose performance on the test was one standard deviation above the mean for his or her age group would have an IQ of 115.

To give you an idea of what an IQ difference of 15 points means, a person with an IQ of 100 might be expected to graduate from high school without much distinction and then attend a year or two of a community college, whereas a person with an IQ of 115 could expect to graduate from college and might go on to become a professional or fairly high-level business manager. In the other direction, someone with an IQ of 85, which is at the bottom of the normal range, is a candidate for being a high school dropout and could expect a career cap of skilled labor.

Although IQ tests were designed to predict school achievement, it quickly became apparent that they were measuring something that overlaps substantially with ordinary people's understanding of what intelligence is. At any rate, people's ratings of other people's intelligence correlate moderately well with the results of IQ tests. Those who are rated higher in intelligence by ordinary people also get higher IQ test scores.

There are a huge number of IQ tests, but there is not much difference among the reasonably comprehensive ones, and the typical correlation between any two tests, even those having rather different apparent content, is in the range of .80 to .90.

Tests of intelligence sometimes measure quite specific skills such as spelling ability and speed of reasoning. Such highly specific tests tend to correlate with one another in clusters. For example, tests of memory tend to be correlated. The same is true for tasks measuring visual and spatial perception of various kinds (for example, arranging colored blocks to match a two-dimensional design) and for tasks measuring verbal knowledge (for example, vocabulary). All tests of anything that you would be likely to call intelligence are correlated at least to some degree. (For that matter, everything that a society holds to be good is correlated with every other good thing to some degree. Life is unfair.)

BOX 1.1 *Subtests Employed on the Wechsler Intelligence Scale for Children (WISC)*

Information:	What continents lie entirely south of the equator?
Vocabulary:	What is the meaning of derogatory?
Comprehension:	Why are streets usually numbered in order?
Similarities:	How are trees and flowers alike?
Arithmetic:	If six oranges cost two dollars, how much do nine oranges cost?
Picture Completion:	Indicate the missing part from the incomplete picture.
Block Design:	Use blocks to replicate a two-color design.
Object Assembly:	Assemble puzzles depicting common objects.
Picture Arrangement:	Re-arrange a set of scrambled pictures so that they describe a meaningful set of events.
Coding:	Match symbols with shapes using a key as a guide.

As an example of a particular IQ test, Box 1.1 shows subtests of the Wechsler Intelligence Scale for Children (WISC), which can be given to children aged six to sixteen. Correlations among subtests of IQ tests like this are in the vicinity of .30 to around .60. That they are correlated is reflected in the notion that there is something that corresponds to general intelligence, a construct known as the *g* factor. (*Factor* has a technical meaning that is unnecessary for us to pursue. The *g* factor is itself highly correlated with

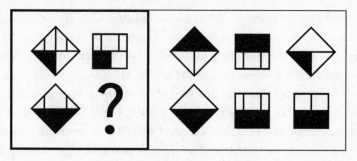

Figure 1.1. A problem similar to those on the Raven Progressive Matrices test. From Flynn (2007). Reprinted with permission.

IQ score, but it is slightly different from IQ score in respects that needn't concern us here.) Some subtests better correlate with the *g* factor than others, and we say that they have high *g* loadings. The Vocabulary subtest, for example, is highly correlated with *g*, whereas Coding (matching symbols using a key) is not.

The Two Types of IQ

There are actually two components to *g* or general intelligence. One is fluid intelligence, or the ability to solve novel, abstract problems—the type requiring mental operations that make relatively little use of the real-world information you have been obtaining over your lifetime. Fluid intelligence is exercised via the operation of so-called executive functions. These include "working memory," "attentional control," and "inhibitory control." The information that you must sustain constantly in your mind in order to solve a problem, and that requires some effort to maintain, is said to be held in working memory. Attentional control is the ability not only to sustain attention to relevant aspects of the problem but also to shift attention when needed to solve the next step in the problem. And inhibitory control is the ability to suppress irrelevant but tempting moves.

Figure 1.1 shows a classic example of a problem that tests fluid intelligence. It's from the Raven Progressive Matrices. The word *matrices* refers to the collection of figures in the problems—arrayed as 2 × 2 or 3 × 3 matrices. The word *progressive* refers to the fact that the problems get harder and harder. John C. Raven published the first version of the test in 1938.

The example that the problem-solver must follow is set up by the two figures in the top row of the left panel. The figure at the left in the bottom row then specifies what has to be transformed in order to solve the problem. The six figures on the right correspond to the answer alternatives. The solution of the problem requires you to notice that the figure in the upper left of the left panel is a diamond and the figure in the upper right is a square. This tells you that the answer has to be a square. Then you must notice that the bottom half of the upper diamond is divided into two, with the left portion in black. The fact that the left half of the figure on the right is also black tells you that the corresponding portion of the square on the bottom right must match the corresponding portion in the lower left diamond—that is, the entire bottom half must be black. Then you notice that to make the top right figure, one of the bars has been removed from the top left diamond while symmetry of the bars has been preserved. This establishes that you must remove one of the bars of the square at the bottom while preserving symmetry. Now you know that the correct answer must be the square at the bottom right of the answer panel.

Of the subtests on the WISC shown in Box 1.1, the ones that most involve fluid intelligence are Picture Completion, which requires you to attend to all aspects of a figure and analyze which portion of it is missing; Block Design, which requires you to operate with purely abstract visual materials; Object Assembly, which requires going back and forth between your knowledge of what the desired object looks like and the abstract shapes that must be used to compose it; Picture Arrangement, which requires you to hold in working memory the various pictures and to rearrange them mentally until a coherent story is told by the pictures in a given order; and Coding, which is a completely abstract task that measures primarily speed of information processing. Scores on these types of tests are sometimes said to comprise the Performance IQ, referring to the fact that all of the subtests require performing operations of some kind. These operations are brought to bear on the spot and draw only somewhat on stored knowledge.

The other type of general intelligence is called crystallized intelligence. This is the store of information that you have about the nature of the world and the learned procedures that help you to make inferences about it. The subtests on the WISC that tap crystallized intelligence most heavily are Information, Vocabulary, Comprehension, Similarities, and Arithmetic. Doing arithmetic, of course, involves both calling up stored or crystallized knowledge and performing operations, most if not all of which have been learned previously. The WISC creators refer to the total of the scores on this collection of subtests as Verbal IQ, since most of the information being drawn upon is verbal in nature. The score that combines Performance IQ with Verbal IQ is called Full Scale IQ.

How do we know that there are two fundamentally different types of general intelligence? We know this first because the subtests that we describe as performance-oriented clearly draw relatively more on reasoning skills (fluid intelligence) than on knowledge (crystallized intelligence), and the subtests that we call verbal clearly depend relatively more on knowledge (including knowledge about algorithmic solutions) than on reasoning skills. Also, the verbal subtests have higher correlations with one another than they do with the performance subtests, and vice versa.

In addition, the subtests that we call measures of fluid intelligence rest on executive functions that are mediated by a portion of the frontal cortex called the prefrontal cortex (PFC) and another region linked in a network with the PFC called the anterior cingulate. The destruction of the PFC has devastating consequences for mental tasks that require the executive functions of working memory, attentional control, and inhibitory control. People with severe damage to the PFC may be so incapable of solving Raven matrices that they function at the level of mentally retarded people on that test, yet have crystallized intelligence that is entirely normal. The opposite pattern also occurs. Autistic children usually have impaired crystallized intelligence but may have normal or even superior fluid intelligence.

As one would expect given the lesion evidence, the PFC is particularly active, as demonstrated in brain-imaging studies, when people attempt to solve problems that make substantial use of fluid intelligence, such as the Raven matrices and difficult mathematical problems.

Additional evidence for the two types of intelligence is that fluid intelligence and crystallized intelligence have quite different trajectories over a lifetime. Figure 1.2 shows an idealized version of those trajectories. The growth of fluid intelligence is rapid over the first years of life but begins to decline quite early. Already by the early twenties, fluid intelligence shows some decline. Mathematicians and others who work with symbolic, abstract materials for which they must invent novel solutions may find their powers fading somewhat by the age of thirty. By seventy years old, fluid intelligence is noticeably less than it has been—more than one standard deviation lower.

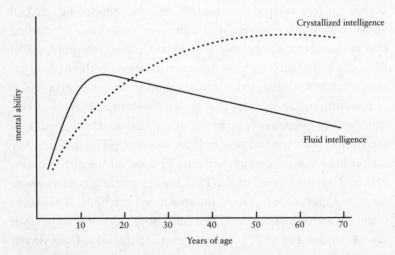

Figure 1.2. Schematic rendering of fluid intelligence and crystallized intelligence over the life span. From Cattell (1987).

Older people find it harder to solve puzzles and mazes. Crystallized intelligence, on the other hand, may continue to increase over the lifespan, at least until very old age. Historians and others whose best work depends on a large storehouse of information may find their powers increasing well into their fifties.

Note that everything I have just written about the age trajectories of fluid and crystallized intelligence is controversial to some degree. I will spare you the ins and outs of this controversy and simply say that the universally agreed-upon fact is that fluid intelligence begins to decline earlier than crystallized intelligence.

That fluid intelligence declines earlier in life than crystallized intelligence could be predicted by the fact that the PFC shows deterioration earlier than other structures in the brain.

A final source of evidence that the two types of intelligence exist is that executive functions and overall IQ may be separately heritable. Executive functions are inherited to a degree from parents, and so is crystallized intelligence, or the knowledge that helps people solve problems. A person can inherit relatively high executive function from parents who have relatively high executive function, yet can inherit relatively low crystallized intelligence from the same parents who score relatively low on this dimension.

Fluid intelligence is more important to good intellectual functioning for younger people than it is for older people. For young children, the correlation between fluid intelligence and reading and math skills is higher than that for crystallized intelligence. In contrast, for older children and adults, the correlation between crystallized intelligence and reading and math skills is higher than that for fluid intelligence. This point becomes crucial later, when I discuss some of the reasons for the relatively low IQ of people of lower socioeconomic status and members of some minority groups, and some of the possible ways to improve IQ.

Another extremely important fact about fluid intelligence is that the PFC has substantial interconnections with the limbic lobe, which is heavily involved in emotion and stress. During emotional arousal, there is less activity in the PFC, and so fluid-intelligence functioning is worse. Over time, continued stress may result in permanently reduced PFC function. This information too becomes critical later, when I discuss the modifiability of fluid-intelligence functioning for the poor and minorities.

In the meantime, however, I focus on the simple IQ score, combining fluid and crystallized intelligence, and make distinctions between the two types of intelligence when that is important.

Varieties of Intelligence

What do IQ scores predict? First of all, they predict academic grades. This is scarcely surprising, because that is why Alfred Binet invented IQ tests more than a hundred years ago. He wanted to be able to determine which children would be unlikely to benefit from normal education and therefore would require special treatment. The correlation today between the scores on typical intelligence tests and the grades of schoolchildren is about .50. That value is substantial, but it leaves room for a great many variables that are not measured by an IQ test to play a role in predicting academic performance.

IQ tests tend to measure what has been called "analytic" intelligence as distinct from "practical" intelligence. Analytic problems typically have been constructed by other people; are clearly defined; have all the information necessary to solve them embedded in their description; have only one right answer; usually can be reached only by one particular strategy; are often not closely related to everyday experience; and are not particularly interesting in their own right. These can be contrasted with "practical" problems, which require recognition that there is something to be solved; are usually not well defined; typically require seeking out information relevant to their solution; have several different possible solutions; are often embedded in everyday experience and require such experience for their solution; and engage—and usually require—intrinsic motivation.

Robert Sternberg measures practical intelligence with questions asking people, for example, how to handle problems like entering a party where one does not know anyone, discussing what share of rental payments is fair for each of several people, and writing a letter of recommendation for someone who is not well known to the person writing it.

Sternberg also writes about a third type of intelligence, which he calls "creative" intelligence. This is the ability to create, invent, or imagine something. He measures creative intelligence by, for example, giving people a title, such as "The Octopus's Sneakers" or "A Fifth Chance," and asking them to write the story. He also measures creative intelligence by asking people to look at a sequence of pictures and having them tell a story about one of them, and by having people develop advertisements for new products.

When Sternberg measures analytic intelligence in the standard way, by SAT or ACT scores or IQ tests, and practical and creative intelligence by his novel measures, he finds that his practical and creative measures add to the predictability of outcomes such as success in school and work performance. Sometimes the increments in predictability are substantial; in fact they sometimes outperform IQ tests by a significant margin.

Sternberg is highly persuasive when he talks about three hypothetical graduate students. Analytic Alice is brilliant in discussion of ideas and is a superb critic of the products of others. Creative Cathy is not so incandescent in her treatment of ideas, but she comes up with lots of interesting notions of her own, a certain fraction of which end up paying off. Practical Patty is neither analytically brilliant nor especially innovative. But she can figure out a way to get the job done. She can get from here to there in sensible, cost-effective ways.

What you hope for are colleagues with all three types of intelligence, of course, but especially when you work on a team, even the people who stand out in only one dimension have a vital role to play. It is worth noting that Sternberg's measures of practical and creative intelligence show much less of a separation between minority and majority groups than do analytic tests, meaning that they become a way to bring more minorities into educational and occupational roles where their entrance might be blocked by tests of analytic intelligence.

Howard Gardner argued that IQ tests measure only linguistic, logical-mathematical, and spatial abilities but neglect other "intelligences." These include various "personal intelligences" resembling the "emotional intelligence" that social psychologist Peter Salovey and his colleagues researched. Emotional intelligence includes being able to accurately perceive emotion, using emotions to facilitate thinking, understanding emotions, and managing emotions in self and others. Emotional intelligence as measured by Salovey and his colleagues is virtually uncorrelated with analytic intelligence as measured by IQ tests, but it predicts peer and supervisor ratings of dimensions like interpersonal sensitivity, sociability, contributing to a positive work environment, stress tolerance, and leadership potential. Some might want to avoid the term *intelligence* for these measures of emotional skills, but that is a quibble.

The other intelligences that Gardner discussed are "musical" and "bodily kinesthetic." Some intelligence researchers are utterly contemptuous of using the term *intelligence* for these skills. But there are such things as musical and kinesthetic ideas, and there are such things as musical and kinesthetic problems to be solved. I'm personally willing to call Beethoven's Seventh Symphony and Alvin Ailey's "Revelations" works of genius. So I am perfectly content to say that the skills that invented those works are the products of intelligence. But I wouldn't try to press my personal preference on others who are resistant to that label.

Gardner justified his lengthening of the list of intelligences by pointing out that there are child prodigies for most of them, and there is neurological evidence that different areas of the brain are specialized for each of the intelligences he identified. Whether one calls his additions to the intelligence list mere skills or something else, it is clear that they are somewhat separate from the standard analytic ones and that measures of them predict—or in principle could predict—important aspects of skilled human endeavor that the standard tests do not.

Motivation and Achievement

Finally, characteristics that no one would call intelligence have a marked effect on academic and occupational achievement.

Decades ago, personality psychologist Walter Mischel studied the ability of children to delay gratification. He put preschoolers from a Stanford University nursery school into a room (alone, they thought, but with an experimenter watching their every move) where there was a cookie, a marshmallow, a toy, or some other desirable object. The children were told that they could have the object whenever they wanted. They had only to ring a bell and then the experimenter would come in and let them have the object. Or they could wait until the experimenter returned of his own accord. If they waited that long they could have two cookies, marshmallows, or toys. The dependent measure is called the "delay of gratification." The longer the child waits before ringing the bell, the greater the ability to delay gratification.

Mischel then waited more than a decade, until these mostly upper-middle-class kids were in high school. The children who had waited the longest before taking the goodie were rated by their parents as being better able to concentrate, better planners, and better able to tolerate frustration and to deal maturely with stress. These traits paid off in tests of measured academic intelligence. The greater the toddler's ability to delay gratification, the higher the SAT score in high school. The correlation between delay time of the toddler and the SAT Verbal score was .42; the correlation for the SAT Math score was .57. It is possible that the brighter toddlers are the ones who had the longer delay times, but it seems unlikely that this is the whole explanation. More plausible is that children who were better able to resist temptation were better able to hit the books when they got older. This is not the last time we will have occasion to note that SAT scores, which correlate highly with IQ scores, are not equivalent to IQ. Some cultural groups do better on SAT scores than would be predicted by their IQ—for reasons that probably have to do with motivation.

That motivational factors affect academic achievement levels is scarcely shocking. That motivation may sometimes actually be a better predictor of academic achievement than IQ is, however, a surprise. This is what an extremely important study found for eighth-graders at a magnet school in a large city in the Northeast. Psychologists Angela Duckworth and Martin Seligman measured self-discipline in a variety of ways. They asked the students about the degree to which they said and did things impulsively; they quizzed the students about different kinds of rewards, asking them about the degree to which they would prefer a small, immediate version of the reward versus a large, delayed reward; they actually offered the children one dollar immediately versus the opportunity to get two dollars a week later; and they asked parents and teachers about each student's ability to inhibit behavior, follow rules, and control impulsive reactions. They combined scores on all of these measures into a single omnibus measure of self-discipline and then compared how well this measure predicted grades versus how well a standard IQ test predicted grades. The result: the IQ test was not nearly as good a predictor of grades as the motivational measure. The correlation for IQ was a very modest .32. The correlation for self-discipline was more than twice as high—.67. The self-discipline measure was a slightly better predictor of standard school achievement test scores than was IQ—.43 versus .36, though the difference was not statistically significant. If you had to choose for your child a high IQ or strong self-discipline, you might be wise to pick the latter.

The results of the Duckworth and Seligman study, important as they are, need to be replicated. The differences between self-discipline and IQ as predictors of achievement might not be the same at a nonselective school or even another magnet school. Nevertheless, the study stands as proof of the hypothesis that motivational factors can count more than IQ as a predictor of achievement.

Let's draw together some of the lessons of the research I have been describing.

IQ is only one component of intelligence. Practical intelligence and creative intelligence are not well assessed by IQ tests, and these types of intelligence add to the predictability of both academic achievement and occupational success. Once we have refined measures of these types of intelligence, we may find that they are as important as the analytic type of intelligence measured by IQ tests.

Intelligence of whatever kind and however measured is only one predictor of academic and occupational success. Emotional skills and self-discipline and quite possibly other factors involving motivation and character are important for both.

To these qualifications of the importance of IQ, we can add the fact that, above a certain level of intelligence, most employers do not seem to be after still more of it. Instead, they claim that they're after strong work ethic, reliability, self-discipline, perseverance, responsibility, communication skills, teamwork ability, and adaptability to change.

So IQ is not the be-all and end-all of intelligence, and intelligence, even when defined more broadly than IQ score, is not the only important factor influencing academic success or occupational attainment. And academic success is itself only one predictor of occupational success.

What IQ Predicts

Nevertheless, IQ and academic success are associated with a great many outcomes. But it is surprisingly difficult to specify exactly what the causal pathways are. Researchers often determine the individual's contemporary IQ or IQ earlier in life, socioeconomic status of the family of origin, living circumstances when the individual was a child, number of siblings, whether the family had a library card, the educational attainment of the individual, and other variables, and put all of them into a multiple-regression

equation predicting adult socioeconomic status or income or social
pathology or whatever. Researchers then report the magnitude of
the contribution of each of the variables in the regression equation, net of all the others (that is, holding constant all the others).
It always turns out that IQ, net of all the other variables, is important to outcomes. But as I make clear in Appendix A on statistics,
the independent variables pose a tangle of causality—with some
causing others in goodness-knows-what ways and some being
caused by unknown variables that have not even been measured.
Higher socioeconomic status of parents is related to educational
attainment of the child, but higher-socioeconomic-status parents
have higher IQs, and this affects both the genes that the child has
and the emphasis that the parents are likely to place on education
and the quality of the parenting with respect to encouragement of
intellectual skills and so on. So statements such as "IQ accounts
for X percent of the variation in occupational attainment" are
built on the shakiest of statistical foundations. What nature hath
joined together, multiple regression cannot put asunder.

But it is possible to get a much firmer bead on the importance
of IQ in determining life outcomes. Political scientist Charles
Murray has looked at people whose IQs were measured by the
Armed Forces Qualification Test as part of the National Longitudinal Survey of Youth begun in the late 1970s. He examined income and other social indicators for the original group
when they were adults many years later. But he did this for a
highly selected sample—sibling pairs who were not born into
poverty (that is, their income was higher than the lowest quartile
of income), who were not illegitimate, and whose parents were
together at least up until the child was seven. But the pairs had to
differ in IQ. Each person either had an IQ in the normal range—
from 90 to 109—or had one outside that range. The sibling with
an IQ outside that range could either be bright (110–119), very
bright (120+), dull (80–89), or very dull (less than 80).

Murray could ask for this sample, which he called "utopian,"
how much difference it made to have IQ in the normal range versus

outside that range in one direction or another. The firmest measure that he had was the income of the individuals as adults. Income, of course, is correlated with occupational attainment and social class, so when we look at income we can treat it as a proxy for these other variables as well. He also had good data on whether the women in the sample had given birth to illegitimate children. This variable too can be taken as a proxy for a large number of other variables, in this case some involving social dysfunction such as likelihood of incarceration and likelihood of being on welfare.

What Murray found is that even in this stable, mostly middle-class group of siblings, different IQs are associated with very different outcomes. Table 1.1 shows that if a person had a normal-IQ sibling but was very bright himself or herself, the very bright sibling made more than a third more money (and thus would have had on average a substantially higher-status job). If a person had a normal-IQ sibling but was himself or herself very dull, the very dull sibling made less than half as much as the normal-IQ sibling. Illegitimate births were also very much tied to IQ. Very dull women were two and a half times as likely to have illegitimate children as their normal-IQ siblings.

TABLE 1.1 *Relationship between IQ and income and percentage of women having illegitimate children, for siblings from the same stable, middle-class family who differ in IQ*

IQ Group	Income	Illegitimacy Rate (%)
Very bright siblings (120+)	$70,700	2
Bright siblings (110–119)	$60,500	10
Reference group (90–109)	$52,700	17
Dull siblings (80–89)	$39,400	33
Very dull siblings (< 80)	$23,600	44

What is important about these analyses is that they show that members of the same family who have different IQs have very different life outcomes on average. Importantly, the analyses remove from consideration the effects of socioeconomic status of the family of origin, since all of the comparisons are between members of the same family. The analyses do not prove that IQ alone is directly causing those outcomes. For example, it is likely that educational opportunity, which is mediated by IQ, is also an important part of the causal chain. In effect, education probably acts as a multiplier of the effect of IQ. And IQ is undoubtedly associated with aspects of character and motivation that play a role as well. But the results are very telling about the importance of IQ and its correlates even among members of the same, relatively high-status and stable, family.

The IQ scores Murray examined are undoubtedly influenced substantially by genes. Some children in a family get a better luck of the genetic draw from their parents. Murray himself has long been associated with the view that IQ is largely genetically determined, and with the view that, partly because of this, IQ is not very susceptible to influence by environmental factors. But how important are genes, exactly? And what role do they leave for the environment? In the next chapter I pursue the questions regarding the degree to which intelligence is something that is inherited and the degree to which the environment can modify it.

CHAPTER TWO

Heritability and Mutability

. . . 75 percent of the variance [in IQ] can be said to be due to genetic variables . . . and 25 percent to environmental variation. —Arthur Jensen (1969)

Being raised in one family, rather than another . . . makes few differences in children's personality and intellectual development. —Sandra Scarr (1992)

NOT SO VERY LONG AGO, scientists who study IQ more or less agreed that intelligence is mostly heritable. Some still consider it to be about 75 to 85 percent heritable, at least for adults. The effects of the environment that children share by virtue of being raised in the same family have often been presumed to be slight, sometimes literally zero by the time the children are adults. Scientists frequently believed, or at least wrote as if they did, that the overwhelming importance of heritability meant that the environment could do little and that social programs intended to improve intelligence were doomed to failure.

But many scientists today consider the heritability of IQ to be much lower than 75 to 85 percent. This environmentalist camp estimates heritability to be .50 or less. (Though, as you will see later, heritability actually differs quite a bit from one population group to another.) And I agree with these scientists—in fact I suspect heritability may be even lower than .50.

In the first section of this chapter, I show why earlier estimates of heritability were so high. Even more important, I review the results of adoption studies showing that raising someone in an upper-middle-class environment versus a lower-class environment is worth 12 to 18 points of IQ—a truly massive effect. This fact places a very high upper bound on the degree to which the environment can influence intelligence. Finally, I emphasize that the heritability of a characteristic places no theoretical limits on the degree to which it can be affected by the environment. The upshot is that the environment counts for a lot in determining IQ and could conceivably account for more if we could think of the right ways to change it.

Some of the notes for this chapter are long. This is because I want to answer the concerns of specialists in the heritability of IQ without derailing the attention and understanding of the general reader. Even without the notes, this chapter is by far the most technical in the book. Please don't get bogged down in it. Rather, take on faith for the time being my assertions that genes are far from being completely determinative of intelligence and that the environment can made a huge difference to intelligence.

Heritability, Environment and IQ

Laypeople sometimes think of the heritability of a characteristic as the degree to which it is inherited from parents. This encourages the assumption that a heritability estimate for IQ of .80 means that 80 percent of a person's IQ comes from genes. This is quite wrong. Heritability does not refer to the individual at all, but to populations. The heritability of a characteristic refers to the percentage of variation in that characteristic in a particular population that is due to genetic factors. This contrasts with the percentage of variation in the characteristic due to all other factors. For intelligence, these other factors include prenatal and perinatal biological factors, environmental factors of a biological nature such as nutrition, and social factors such as education and experience. This chapter concerns the most interesting sources of variation—that due to genetics and that due to the environment which is shared among children in a given family but differs between families.

The between-family environmental effect refers to how much difference it makes that a person was raised in one family versus another (with all the various factors that go with membership in different families, such as social class, rearing styles, and religious orientation). The between-family environmental source of variation does not include environmental variation *within* a family, such as that associated with birth order. Only one child in a family can be the first born, only one can be the second born, and so on. And we know that birth order can be an important factor associated with some characteristics. Other factors such as peer influences and schools attended also differ among children within a given family.

Of course, between-family environmental differences do not include the genetic contribution of the parents. Everyone agrees that there are big average differences in IQ between two randomly selected families, and a substantial portion of those differences is due to genes.

The researchers I call the strong hereditarians hold that IQ is between 75 and 85 percent heritable in the population of developed countries, and that the environmental contribution from all sources is between 15 and 25 percent. Most of the strong hereditarians believe that the between-family environmental contribution (growing up with the Smiths rather than the Joneses) is slight to nil, at least after childhood. They also agree that such environmental contribution as exists is mostly due to variations that occur within the family—for example, the children having gone to different schools, having been treated differently by their parents, and having experienced a different uterine environment.

How do the strong hereditarians reach these conclusions? Have a look at Table 2.1, which summarizes the results of a large number of studies of the correlations between individuals with a given degree of relatedness who were raised either together or apart. A direct estimate of heritability is provided by the figure for identical twins—who have the same genetic makeup—who are reared apart. This figure is .74, and is essentially the one that Arthur Jensen drew upon to derive his estimate of the contribution of genes to IQ. Since the environments of the twins are different, so the logic goes, the similarity between the twins can only be due to genetics (as well as to prenatal and perinatal effects that might have been consequential before the twins were separated and which are generally assumed by the hereditarians to be minimal).

TABLE 2.1 *Correlations for individuals with varying degrees of relationship, raised either together or apart*

Relationship	Raised	Correlation
Identical twins	Together	.85
Identical twins	Apart	.74
Fraternal twins	Together	.59
Siblings	Together	.46
Siblings	Apart	.24
Midparent/child	Together	.50
Single parent/child	Together	.41
Single parent/child	Apart	.24
Adopting parent/child	Together	.20
Adopted children	Together	.26

Midparent = average of mother's and father's IQ.

Correlations are based on a summary of 212 different studies and are weighted by the size of the sample. From Devlin, Daniels, and Roeder (1997), except for the correlation for adopted children raised together, which is from Bouchard and McGue (2003).

How do scientists typically arrive at a direct estimate of the role of between-family environmental differences? They examine the correlation between the IQs of unrelated individuals living together. A direct way to estimate this is by determining the correlation between the IQs of adopted children and those of their adoptive parents. Since these people do not share genes, the only way that the IQs of the adopted children can resemble those of their adoptive parents is by virtue of sharing the same environment. The next-to-last line of Table 2.1 shows that this correlation is .20. Some scientists take this as a good estimate of the contribution of the environment to the variation in IQ from one family to another. A different way of arriving at this conclusion is to compare the IQs of unrelated children living in the same family (the last line in Table 2.1). Again, because the children do not share genes, the only way they can resemble one another is through sharing the same environment. This correlation is .26, a slightly higher estimate of the contribution of between-family differences to IQ.

Jensen and other strong hereditarians, however, would not accept a figure for between-family environmental effects that is as high as .20 to .26. This is because when people older than those in the studies summarized in Table 2.1, who are mostly children, are examined, the correlations drop dramatically—sometimes to as low as zero. This is true, for example, for unrelated children brought up in the same household. When they are adults, the correlations run in the vicinity of .05 or less. The usual explanation given for this weak effect on adults is that as people grow older, they select their own environments, and their preference for one environment versus another is largely influenced by genetics. The importance of the early environment, never all that great to begin with, drops way off. This means that the strong hereditarians assign most of the environmental contribution to IQ to factors that differ among members of the same family, such as birth order, rather than to factors that are common to the members of the same family and that differ between families.

To summarize, the strong-hereditarian position is as follows: three-quarters or more of the variation in IQ is genetic; some of the variation in IQ is due to nonshared, within-family environmental factors that a parent cannot do much about; and by adulthood almost nothing of the variation in IQ is due to between-family environmental differences—the difference between randomly chosen family A and randomly chosen family B. So the characteristics of your family, in comparison to the characteristics of the randomly selected Joneses—who might make less money, not read as much to their kids, send them to poorer schools, live in a sketchier neighborhood, and have a different religion—make almost no difference.

By now, if you have children, you could be wondering why you spent good money to move to a more expensive neighborhood with better schools, or for that matter why you squander money on books and orthodontia, waste time driving them to violin lessons and museums, and drain off emotional energy holding your temper so as to set a good example. But you don't have to accept those high estimates of heritability and low estimates of between-family environmental effects.

The direct estimate of heritability based on the correlation between the IQs of identical twins reared apart makes a tacit assumption that is surely false—namely, that the twins were placed into environments at random. For that to be true, the twins would have to be in environments that differed as much as any two people selected at random from the Google-based U.S. telephone book. But that is not the case. Billy is likely to be raised by people—frequently relatives in fact—not all that different from those raising Bobby. And just how similar the environments are makes a good deal of difference for the correlations in IQ between identical twins reared apart. Developmental psychologist Urie Bronfenbrenner showed that when twins reared apart are brought up in highly similar environments, the correlations between their IQ scores range from .83 to .91. But the correlation now reflects not just the fact that their genes are identical but also the fact that their environments are highly similar. Such a correlation therefore gives an inflated estimate of heritability. When environments are dissimilar to one degree or another, the correlations range from .26 to .67. Since we don't know just how dissimilar the environments are in most studies of twins reared apart, we can't know exactly what heritability to estimate from the correlation between them.

Regardless of the degree of similarity in environments, the correlation between identical twins overestimates heritability as compared to other ways of estimating heritability based on the correlations for other kinds of relatives. This could be because the environmental experiences of identical twins who are reared separately in quite different environments are highly similar since they look so much alike or have other characteristics in common that tend to elicit the same sorts of behavior from other people. Or there may be gene interactions that specifically make identical twins more similar but that don't contribute much to the degree of resemblance of other relatives.

A third source of error in the .75 to .85 estimates is that twins share the same uterine environment. Devlin and his colleagues maintain that this shared-environment factor means that as much as 20 percent should be subtracted from the heritability estimates.

I will present a fourth source of error later, when I discuss the fact that heritability differs greatly from one social class to another—and that twin studies are biased toward including disproportionately large numbers of people from social classes for which heritability is high.

Once corrections are made for all these facts, the estimate based on identical twin correlations is likely to be substantially lower than the .75 to .85 assumed by Jensen and the other strong hereditarians.

Genes as Triggers of Environmental Influences

Developmental psychologists Sandra Scarr and Kathleen McCartney, as well as economist William Dickens and philosopher and IQ scientist James Flynn, propose a further reason why the role of genes is overestimated. Slight genetic advantages can be parlayed into very great IQ advantages because of the way they influence the kinds of experiences an individual has. Consider a basketball analogy. The child who is somewhat taller than average is more likely to play basketball, more likely to enjoy the game, more likely to play a lot, more likely to get noticed by coaches and encouraged to play for a team, and so on. That height advantage is totally dependent on such environmental events for its expression. And identical twins reared apart are likely to have very similar basketball experiences, because they are of similar height, so they are likely to end up with similar skills in basketball. But their similarity in basketball skills is not due to their possessing identical "basketball-playing genes." Instead, it is due to genetic identity in a narrower attribute that causes them to have highly similar basketball-related experiences.

A similar point can be made about intelligence. A child with a relatively small genetic advantage in, say, curiosity is more likely to be encouraged by parents and teachers to pursue intellectual goals, more likely to find intellectual activity rewarding, and more likely to study and engage in other mental exercises. This will make the child smarter than a child with less of a genetic advantage—but the genetic advantage can be very slight and can produce its effects by virtue of triggering environmental "multipliers," which are crucial for realizing that advantage. All of this gene-environment interaction (or as geneticists would prefer to say, gene-environment correlation), however, gets credited to heritability, given the way heritability is calculated. This is not wrong, but it leads to an underestimation of the role of the environment.

To make it even clearer why heritability estimates slight the role of the environment, let's return to the basketball analogy. Suppose a child of average height is encouraged to play basketball, perhaps because her older siblings are players and keep a well-used hoop on the driveway. And suppose another child, of above-average height, has little access to basketball experiences, perhaps because she lives in a rural area and there are no neighborhood kids nearby. The taller child from the rural area is not likely to become much of a basketball player, whereas the average-height child has a reasonable chance of becoming good. Now we have a child with a genetic advantage who is not very good at basketball, and a child with a genetic disadvantage who nevertheless is pretty good at it. Genes count, and given a constant environment they may have a big influence in determining talent. But environmental interventions can greatly influence—even largely override—the effects of genes. This is particularly important for estimates of the effects of the environment on IQ. It is easy to imagine any number of ways in which intellectual pursuits can be made more or less available and attractive.

Tolstoy and Adoption

Now let's consider how the effects of between-family environmental differences are measured. These differences are estimated by calculating the correlation between the IQs of adopted children and those of their adoptive parents, and the correlation between children in the same family who are not related to each other (usually adopted children). As we have seen, the correlations turn out to be low on average—around .20 to .25. But these numbers only make sense if we assume that the variation in the environments created by adoptive parents is about the same as the variation in the population as a whole. It turns out, though, that adoptive families, like Tolstoy's happy families, are all alike.

Psychologist Mike Stoolmiller showed that the variation in factors that predict IQ is, for adoptive families, a fraction of what it is for families in general. We know this for two reasons. First, the socioeconomic status (SES) of adoptive families is higher than that of nonadoptive families; the bottom rung of the SES ladder is scarcely represented among adoptive families. Second, there is much less variation in scores derived from a method of assessing home environments called the HOME (Home Observation for Measurement of the Environment). HOME researchers assess family environments for the amount of intellectual stimulation present, as indicated by how much the parent talks to the child, how much access there is to books and computers, degree of warmth versus punitiveness of parents' behavior, and so on. HOME assessments show that adoptive families rate far above the general run of families in these respects. In fact, adoptive families score at the 70th percentile on average. Just as important, the range on these variables is very restricted compared to the population at large. HOME measures for disadvantaged families are five times as variable as those for adoptive families; that is to say, disadvantaged families differ far more from one another on average than do adoptive families.

Why does restricted range of environmental variation result in correlation estimates that are too low? Because if there is very little variation in one variable being correlated with another, the correlation cannot be very high. Consider the extreme case of variable A, which has no variation: the correlation with variable B is zero. High and low scores on B would alike be associated with the same score on variable A, so the correlation between scores on variables A and B can only be zero. Therefore, if the variation in environments between adoptive families is mistakenly estimated as being higher than it really is, the impact of the environment on IQ will be underestimated.

Because the environmental variation of adoptive families has mistakenly been assumed to be as great as the environmental variation in the population as a whole, the estimates of between-family environment effects are way off. Stoolmiller calculated that if you correct for this restriction of environmental range, as much as 50 percent of the variation in intelligence could be due to differences between family environments. Since we know that within-family variation also makes an important contribution to IQ, this would mean that most of the variation in IQ is due to the environment. (These numbers would hold, though, only for children. We know that heritability goes up with age to some degree, so Stoolmiller's estimate for the contribution of between-family differences has to be lowered by some unknown amount.)

There Is No Such Thing as the Heritability of IQ (or Anything Else, for That Matter)

So what value should be assigned to heredity's contribution to IQ? Actually, geneticists say that there is no such thing as a single point estimate for heritability. Heritability is dependent on the particular population and the particular circumstances in which it is examined. For IQ in particular, the nature of the population turns out to be crucial. Psychologist Eric Turkheimer and his colleagues recently showed that heritability is radically dependent on social class. They found that the heritability of IQ was about .70 for children whose parents were upper-middle class but was about .10 for children whose parents were of lower social class. A plausible explanation for this is that higher-SES families are providing excellent conditions for the development of intelligence, and they may not differ much among themslves in these respects. Under these circumstances, the contribution of heredity can be very great. At the extreme, if the environment is completely constant across families, the only possible source of variation is genetic.

Why should the heritability of IQ be so low for lower-SES people? We know from the work by Stoolmiller that the range of environments with respect to the variables that influence IQ is far greater for lower-class families than for middle- and upper-middle-class families. The environment for lower-SES people probably varies from being equal to the most supportive upper-middle-class setting to being pathological in every respect. This means that the environment for this group of people is going to make a great deal of difference to IQ. And in fact the environment almost completely swamps heredity.

So you haven't wasted your time, money, and patience on your children after all. If you were to average the contribution of genetics to IQ over different social classes, you would probably find 50 percent to be the maximum contribution of genetics. Most of the remainder of variation in IQ is due to environmental factors— those shared within families and differing between families, plus those not shared within families. (The rest—a small amount—is due to measurement error.)

Note also that the Turkheimer findings provide yet another, and crucially important, indication that the very high estimates of heritability are overestimates. This is because they are based largely on twin studies, and there is a substantial bias in twin studies toward middle-class participants since middle-class and upper-middle-class people are easy to contact and to persuade to participate in research projects. Therefore, estimates of heritability for adults are biased upward, and estimates of between-family environmental effects correspondingly biased downward.

Stoolmiller's claims about the superiority of adoptive parents from the standpoint of encouraging the development of intelligence raise the question of just how big the effects of adoption are. Addressing this issue would provide another way of drawing a bead on the contribution of family environment to IQ. If adoptive families are such an exemplary bunch with respect to the variables that predict IQ, shouldn't we find adopted children to have IQs that are higher than would be expected from their origins? We certainly would under the hypothesis that environments matter a great deal to IQ.

The Proof of the Importance of Family Environment in Determining IQ

Heritability estimates are based on correlations, and as we have just seen, inferences based on correlations can be misleading. What is needed to test the power of differences between family environments are experiments. As it turns out, there are a large number of natural experiments based on the everyday occurrence of adoption. We can ask if it makes a difference whether children are adopted into a family with highly favorable environmental conditions versus a family with less favorable conditions. Many natural experiments, with many different "designs," point to the same conclusion: being raised under conditions highly favorable to intelligence has a huge effect on IQ.

Psychologists Christiane Capron and Michel Duyme carried out a "cross-fostering" study with French children. They tracked down children born to low- or high-SES parents who had been adopted by either low- or high-SES parents. The class differences were pronounced: they compared children of poor and working-class parents based on occupation of the father (semi-skilled or

unskilled worker having nine years of education or less) with children of upper-middle-class parents (professional or upper-level
manager, having an average of sixteen years of education). This
design allowed an independent assessment of the contribution to
IQ of genes from parents of very low versus very high SES and the
contribution to IQ of being raised in a very-low- versus very-high-
SES family. As it turns out, both genes and class-related environmental effects are powerful contributors to intelligence.

On average, the biological children of high-SES parents had IQs
that were 12 points higher than those of low-SES parents, regardless of whether they were raised by high-SES parents or low-SES
parents. (We don't know how much of this difference was due to
genes and how much was due to nongenetic prenatal, perinatal,
and immediate postnatal environmental factors, though I don't
doubt that most of the difference was genetic.)

The crucial finding is that children adopted by high-SES parents had IQs that averaged 12 points higher than the IQs of
those adopted by low-SES parents—and this was true whether
the biological mothers of the children were of low or high SES.
So the study showed that being raised in higher-social-class environments produces children with a far higher IQ than does being
raised in lower-social-class environments. Equally important, the
school achievement of children raised in upper-middle-class environments was much higher than that of children raised in lower-
class environments.

Another French study, having a different "natural experiment
design," examined lower-SES children adopted into upper-middle-
class families and compared them with their siblings who had not
been adopted. The adopted children had an average IQ of 107 by
one test and 111 by another, whereas their biological siblings who
were not adopted had an average IQ of about 95 by both tests.
We therefore get an estimate of 12 to 16 IQ points as the value
of being raised in an upper-middle-class environment versus a
lower-class environment. The difference in academic achievement
between adopted and nonadopted siblings was huge. The school

failure rate was 13 percent for children adopted by upper-middle-class parents and 56 percent for their nonadopted siblings.

In another extremely important natural experiment conducted in France, but with yet a different design, Duyme and his colleagues looked at abused, low-IQ children who were tested for IQ adopted at the age of four or five and then retested for IQ at age fourteen. They deliberately searched for children who had been adopted into families of varying social classes. When younger, the children had IQs between 61 and 85—between what IQ testers describe as decidedly retarded and dull-normal levels. The families into which they were adopted were poor (unskilled workers), lower middle and middle class (lower- or middle-level managers, tradesmen, and skilled laborers), or upper middle class (professionals and high-level managers). The effects on IQ of being adopted were very large, 14 points on average. But the social class of the family made a great difference. Children adopted into lower-SES families gained 8 points; those adopted into middle-class families, 16 points; and those adopted into upper-middle-class families, almost 20 points. This gives an estimate of 12 points for the effect of being raised by an upper-middle-class family as opposed to a lower-class family.

Conveniently, from the standpoint of being able to make a confident inference about the effects of the social class of adoptive parents, there was no selective placement of children. That is, relatively lower-IQ children were no more likely to be adopted into lower-SES families than were relatively higher-IQ children. So this study reached the same conclusions as the other two French studies: the difference between being raised in a lower-class environment versus an upper-middle-class environment is 12+ points. Note that the study showed that being raised in a relatively modest lower-middle or middle-class family can give a big boost in IQ over being raised in a lower-SES family, namely, 8 points. Note also that this study gave, if anything, an underestimate of the effects of being raised in a higher- versus a lower-SES family, because, as Stoolmiller showed, even lower-SES families who adopt have parenting practices of a kind that encourage the growth of intelligence.

A review that examined all of the well-designed adoption studies available as of 2005 found that the effect on children's IQ of being adopted by a middle- or upper-middle-class family as opposed to being left behind in the family of origin (which was generally of lower SES) is 1.17 SD, which translates into an 18-point advantage for upper-middle- versus lower-class upbringing. The review also gave an estimate for the contribution of biological factors—genetic plus prenatal, perinatal, and immediate postnatal factors. This estimate was derived by comparing the biological children of middle- and upper-middle-class families with their adopted siblings. As it happens, the difference was 12 points, the same value Capron and Duyme found.

The crucial implication of these findings is that the low IQs expected for children born to lower-class parents can be greatly increased if their environment is sufficiently rich cognitively.

The review of adoption studies tells a somewhat less optimistic story for school achievement. The adopted children performed only .55 SD better in academic achievement measures than their siblings who were not adopted. On the other hand, they scored only about .25 SD below the general population and even less behind when compared to their classmates.

Before leaving the topic of adoption, I would like to point out that it is customary for strong hereditarians to maintain that the primacy of genes and the low relevance of the environment are established by the fact that there is typically a much higher correlation between biological parents' IQ and their offsprings' IQ than there is between adoptive parents' IQ and their adopted children's IQ. The hereditarians believe that the environment of the adopted child does little for the child's intelligence, since differences in adoptive environments do not make for differences in IQ. We can see now how mistaken this conclusion is. The environments of adoptive families are highly similar for the most part, chiefly consisting of stable middle- and upper-middle-class families. Even adoptive families who are of lower SES are high on the parenting practice factors that predict a high IQ. Since variation in adoptive families is relatively slight, very high correlations would not be expected between the IQs of adoptive parents and those of their children. There just isn't that much difference between the environments of adoptive parents on the dimensions that matter for determining IQ—and if the differences are small, the correlation cannot be big. But there is a huge difference between the adoptive environments on the one hand and lower-SES environments in general on the other, and this results in big IQ differences. So the relatively low correlation between the IQs of adoptive parents and those of their adopted children is nothing but a red herring—it does not contradict in the least the fact that adoptive families are having a huge effect on their adopted children's IQs.

Finally, Herrnstein and Murray state in *The Bell Curve* that the "consensus" about the average effect of adoption on IQ is 6 points, but their evidence for that was a review by Charles Locurto. Locurto reported that the average effect of adoption was 12 points when working-class upbringing was compared with upper-middle-class upbringing.

The belief that differences between family environments have little effect on IQ has to be one of the most unusual notions ever accepted by highly intelligent people. Judith Rich Harris, the author of the very interesting and best-selling book *The Nurture Assumption,* premised her work on the assumption that the contribution made by differences between families is virtually nil. In his brilliant book *The Blank Slate,* Steven Pinker insisted on the same principle. In the best-selling *Freakonomics,* Steven Levitt and Stephen Dubner were explicit that adoption has little effect on intelligence: "Studies have shown that a child's academic abilities are far more influenced by the IQs of his biological parents than the IQs of his adoptive parents." (I wish I could exempt myself from the company of strange believers, but unfortunately for many years I bought—but was deeply puzzled by—the claims of the hereditarians that family environments do not matter much.)

The evidence we have just been looking at concerning the effects of genes versus the environment tells us something crucially important about social class and intelligence. The experiences of the children of the professional and middle classes result in much higher IQs and much lower school-failure rates than is typical for lower-SES children. Moreover, we can place a number, or at least a range, on the degree to which environmental factors characteristic of lower-SES families reduce IQ below its potential: it is between 12 and 18 points. Whatever the estimates of heritability turn out to be, nothing is going to change this fact. So we know that, in principle, interventions have the potential to be highly effective in changing the intelligence of the poor. Interventions could also greatly affect the rate of school failure of lower-class children. The minimum estimate for this reduction is about half a standard deviation. The maximum estimate for this is much higher—one standard deviation, or about the same rate that would be found for middle-class children raised by their own parents.

Note also that it is not just the IQs of lower-SES children that can be affected. One study looked at the IQs of white children who were born to mothers with an average IQ and who were adopted by mostly middle- and upper-middle-class families. The children adopted relatively late had an average IQ in childhood of 112 and those adopted relatively early had an average IQ of 117. This study suggests that even children who would be expected to have an average IQ if raised in an average environment can have their IQ boosted very considerably if they are raised under highly propitious circumstances. Similarly, the cross-fostering study of Capron and Duyme showed that upper-middle-class children can have their IQs lowered if they are raised in poverty. The loss is about 12 points. So children born to poor families are not the only ones who can have their IQs dramatically affected by the environment. All children can.

Heritability Says Nothing about Mutability

Now I can deal a final blow to the idea that high heritability of IQ means that the environment has little effect. The degree of heritability of IQ places no constraint on the degree of modifi-ability that is possible. This is so important that I need to say it again, more emphatically: *the degree of heritability of IQ places no constraint on the degree of modifiability that is possible.* All geneticists accept this principle, but hereditarians often acknowl-edge the principle and then go on to write as if heritability does in fact place limits on modifiability.

To understand why heritability implies nothing about mutabil-ity, think about two facts: (a) the heritability of height is about .85 to .90 and (b) gains in average height of a standard deviation or more have appeared in a generation or less in several countries in the world. The average height of thirteen-year-old Korean boys increased by more than seven inches between 1965 and 2005, a difference of 2.40 SDs. The average boy in 1965 would have been painfully short in 2005. The forty-year time span is far too brief for genetics to have played a role in the increase. The increases in height we have seen in many places in the world in the last couple of generations are obviously due to environmental changes of some kind, probably in nutrition.

Or we can think of an even more extreme case: heritability of 1.0 yet massive environmental influence. We can randomly toss the seeds for corn plants into either rich soil or poor soil. Imagine that the heritability of the height of corn plants is 1.0 in rich soil and 1.0 in poor soil. The average height of the two groups of plants can nevertheless be greatly different and will be entirely due to environmental factors.

These examples should make it clear that the heritability of a characteristic within a given population places no theoretical con-straint on the modifiability of that characteristic by environmental forces. And that's good, because, as you will see in the next chap-ter, despite the moderate heritability of IQ, it is enormously influ-enced by environmental factors—namely, school and the social changes that have taken place over the last eighty years.

CHAPTER THREE

Getting Smarter

> *. . . even a perfect education system is not going to make much difference in the performance of children in the lower half of the distribution . . .* —Charles Murray (2007)

> *. . . a person's total score [on the Raven Progressive Matrices test] provides an index of his intellectual capacity whatever his nationality or education.*
>
> —Raven, Court, and Raven (1975)

THE EXTREME HEREDITARIAN VIEW holds that nothing in the environment can much affect intelligence. You have the IQ your genes had planned for you. This view has two important implications: school should not much affect intelligence; and the intelligence of the population as a whole cannot change a great deal, short of genetic engineering.

That such important predictions can be shown to be so manifestly wrong does not happen often in the behavioral sciences.

Does School Make You Smarter?

Some psychologists have been quite explicit in maintaining that education is mostly irrelevant to intelligence. They hold that education teaches specific facts and procedures but it does not improve raw problem-solving ability to deal with unfamiliar situations. Reflecting the opinion of many intelligence theorists, Herrnstein and Murray, in *The Bell Curve,* acknowledge that more schooling is associated with higher IQs, but they maintained that the association is in large part due to smarter people deciding to stay in school longer. Smarter people like school better and receive more reinforcement for staying in school and so they get more education. This conclusion is based on regression analyses, which, as I pointed out in Chapter 1, are usually unable to answer questions about causality. What happens when education is withheld? Does this prevent people from being as smart as they would otherwise be? Experiments addressing these issues have been tried many times, and the results always turn out to be the same.

Evil social scientists do not conduct these studies. Rather the experiments are natural ones in which children are deprived of school for a period of time for a variety of different reasons. Such experiments eliminate the possibility that more schooling is associated with higher IQ simply because smarter people prefer to remain in school longer. Developmental psychologists Stephen Ceci and Wendy Williams have described these studies at length.

One natural experiment is summer vacation. Kids are deprived of school over the summer, and this results in a drop or greatly reduced growth in IQ and academic skills. The summer slump is especially great for math, for children in the higher grades, and for children of lower socioeconomic status. Much, if not most, of the gap in academic achievement between lower- and higher-SES children, in fact, is due to the greater summer slump for lower-SES children.

The very oldest study on the effect of schooling on IQ was carried out in 1923. Psychologist Hugh Gordon studied the IQ of the children of transients in London, such as canal boat pilots and gypsies, who went to school rarely if at all. The children's IQ was in the low-normal range around the time they would normally start school, but showed a steady decline thereafter. Children four to six years old had an average IQ of about 90—toward the bottom of the normal range—whereas the oldest children (twelve to twenty-two) had an average IQ of about 60, well below the cutoff for mental retardation. The average IQ of children who attend school doesn't drop. So the study findings indicate that schooling is necessary for children to maintain their intelligence.

Another early natural experiment arose from the fact that at the beginning of the twentieth century, some children in remote areas of the United States received little or no education. In the "hollows" around the Blue Ridge Mountains, there were children of the original Scotch-Irish and English immigrants whose forebears had departed to the remote highlands when their land was deeded to German immigrants in the nineteenth century. Most of these children had little access to schools, newspapers, or movies. The older the children were, the lower their IQs were, as tested by performance measures (such as a block design test) that do not require literacy. But children in one of these communities did have a reasonable amount of schooling, and their IQs did not drop as they got older.

World War II provided another natural experiment. School was delayed for Dutch children for several years by the Nazi siege. The average IQ for these children was 7 points lower than for children who came of school age after the siege.

Children of Indian ancestry in many South African villages in the mid-twentieth century had their schooling delayed for up to four years because there were no teachers in their village. The IQs of such children, compared to those of children in neighboring villages with access to school, averaged 5 points lower for every *year* that their schooling was delayed. Even after several years in school, the school-deprived children still had not caught up.

Another late-start-schooling study examined black children in Prince Edward County in Virginia, which shut down its public schools between 1959 and 1964 in order to avoid racial integration. The IQs of the children who had no school during that period diminished by an average of 6 points per year of school missed.

Dropping out of school early is also bad for intelligence. Two different groups of Swedish psychologists obtained the IQs of thousands of randomly selected boys who had taken an intelligence test at the age of thirteen. They equated the children for IQ, socioeconomic status (SES), and school grades at age thirteen and then looked at their IQs again when all the boys had to take another intelligence test to register for the military at age eighteen. The psychologists found that for every year of education skipped, there was a loss of about 2 points for boys having the same IQ, SES, and school grades at age thirteen. The loss for boys who dropped out of school four years early was 8 points—equivalent to half a standard deviation. Note that these studies established that it isn't just that smarter kids stay in school longer and therefore end up smarter, but that regardless of what the IQ was at age thirteen, big gains resulted from staying in school. (Or, alternatively, big losses resulted from dropping out. Given the fact that average IQ is forced to be 100, we cannot be sure whether there was a loss for the dropouts, a gain for those who stayed, or some of both.)

The fact that children can start school in a particular year only if they are born by a certain date allows for an ingenious way of establishing that school makes children smart. For example, many districts have a cutoff date in September. To illustrate, let's pick the date September 15. A child born on September 16 has to wait a year longer than a child born on September 15 before she is allowed to start kindergarten. This makes it possible to examine the results of a very tidy natural experiment. We can compare the IQs of kids who have the advantage of being almost a year older than other kids with the IQs of kids who have the advantage of a year's extra schooling. We can then see which is more important: a year of age or a year of school. Sir Cyril Burt and the Ravens and their modern followers are quite clear in their predictions: a year of age for a young child should be worth a lot, and a year of school should be worth little or nothing. (To be exact, the Ravens would say that a year of school should be worth nothing for an IQ test such as the Raven Progressive Matrices, which allegedly measures pure fluid intelligence uncontaminated by culture.)

In fact, studies in Germany and Israel discovered that a year of school is worth about *twice as much* as a year of age.

Western-style education can have big effects on the IQs of children who previously had only non-Western schooling or none at all. Western-style schools improve memory, including memory of the type that IQ specialists often claim is influenced little or not at all by academic learning, such as that assessed by digit span (ability to remember digits that are presented by voice) and coding tests (where the child matches symbols with shapes or numbers using a key). As little as three months of a Western-style education improved the ability of African teenagers to perform a variety of spatial perception tasks of the kind found on IQ tests by as much as .70 standard deviation. These included performance tests such as block design, memory for designs, and picture description—tasks that IQ researchers have often characterized as being measures of raw, unschooled intelligence.

Given that schools directly teach material that appears on comprehensive IQ tests, including information such as the name of the writer who wrote *Hamlet* and the elements that make up water, as well as vocabulary words and arithmetical operations, it is strange that some IQ theorists doubt that school makes people more intelligent. What is still more surprising to traditional IQ theorists is that school also affects people's ability to solve problems that have been regarded as culture-free, such as those on the Raven Progressive Matrices test. Everyone has had experience with circles and squares and triangles, so many IQ theorists assume that such completely abstract tests are free of the influence of schooling. As you'll now see, it turns out that this assumption is very far off the mark.

Are We Smarter than Our Grandparents?

Given that school makes us smart, and given that we have much more education now than people did a hundred years ago, wouldn't it seem to follow that we are smarter than our great-grandparents were? In America in 1900, the mean level of schooling completed was seven years, and a quarter of the population had finished four years or less. The mean today is two years of education after high school, or fourteen years, and the great majority of people complete high school.

If you are familiar with the fact that the average IQ has been 100 for almost a century, you might be inclined to assume that education has had no impact on intelligence. But IQ tests are designed to give an average of 100 by definition, so the constant 100 average actually reveals nothing about change in intelligence over time. To find out whether people get better scores on IQ tests, you would give them the tests that people of a previous time period took, and compare the performance of the group tested earlier with that of the group tested later. This is what happens when tests are re-normed. If the same test were to be given year after year, IQs would become ever higher. To keep the average IQ at 100, new, more difficult items are added to the test.

So people are actually doing better with each succeeding year on the kinds of skills that IQ tests capture. For major IQ tests such as the Wechsler Intelligence Scale for Children (WISC), the Wechsler Adult Intelligence Scale, and the Stanford-Binet, the gain has been almost a third of a point per year in the fifty-five-year period between 1947 and 2002. This amounts to a total of 9 points per thirty-year generation in the United States. James R. Flynn has documented this increase, which has been christened (not by him) the Flynn effect. The rapid increase in IQ has been found in all developed nations where change has been investigated. In some countries it has been somewhat smaller than in the United States, and in others, somewhat larger.

What is responsible for this amazing increase? In what follows I stick closely to the most recent account in an important book by Flynn.

One guess about why the gain has occurred is that it might be due to increasing test-wiseness—greater familiarity with standard-ized, paper-and-pencil tests. This explanation is unlikely. The IQ gain has been going on since at least 1917. Scores for eighteen-year-olds on the army's IQ test went up by 12 to 14 points between then and the start of the draft for World War II, and people would have had little exposure to standardized tests during that period. The gain in IQ has been more or less constant, covering a period from when relatively few people had experience with standardized tests, to more recent decades, when everyone has had innumer-able exposures to them. And in any case, if completely test-naive people are given actual IQ tests repeatedly, their IQ scores do go up somewhat, but not a lot. And we are looking at an 18-point gain in the period between 1947 and 2002.

How about nutrition? This is also quite unlikely. While poor nutrition undoubtedly negatively affects IQ in some parts of the world today, and may even have done so in the United States and Europe before World War II, there is little evidence today of poor nutrition of a magnitude that would likely stunt mental growth for very many people. Most nutrition deficits in developed countries today occur in the prenatal or immediate postnatal period, and while this may have lessened in recent decades, it has been argued that the net effect on IQ of the population has probably been neutral. For every child whose intelligence prospects are improved by better perinatal nutrition, another child has been saved from death but is nevertheless mentally impaired. In any case, in all probability very few children in developed countries have been affected by poor nutrition in recent decades.

An additional argument against a role for nutrition is that scores have increased evenly throughout the range for IQ. The IQs of people at the top third of the distribution, for whom there would not have been nutritional disadvantages in any recent era, have increased as much as the IQs of people at the bottom third. The even gains across the distribution of IQ, incidentally, establish that Charles Murray is wrong when he says that nothing much can be done to improve the IQs of people in the bottom half of the distribution.

But then what does the gain show? We have to grapple with the meaning of an 18-point increase over fifty-five years and a rate of gain of about the same or even more in the preceding thirty-year period. Suppose we take as the real value of intelligence in 1947 an IQ of 100. Typical jobs for a person with an IQ of 100 would include skilled worker, office worker with a not terribly responsible job, and salesclerk. A four-year college would have been very difficult for such a person to handle, even if the person had the means to go to one. The average grandchild of that average person would have an IQ of 118 using the same test. A person with an IQ of 118 is not only capable of doing excellent college work but is likely to be able to do postgraduate work if that's desired and to become a professional such as a doctor or a lawyer, a high-level manager, or a successful entrepreneur. Is it possible that people have gotten that much smarter on average?

Or let's work the numbers back the other way. Say that the true value of an average person's IQ in 2002 was 100. The average grandparent of that average person would have had an IQ of 82 using the version of the test given today to the grandchild. That grandparent would not likely have been able to carry out an office job of much responsibility and would have likely been challenged by the requirements of most skilled labor. Completion of high school would have been an iffy bet.

Or let's project back another thirty years to 1917. The great-grandfather of today's average person would be expected to have had an IQ of 73 using today's test! Skilled labor would have been unlikely for that great-grandparent; high school completion would have been out of the question. And half the population would have been considered retarded by today's standard!

Something is clearly desperately wrong with this picture. We are not that smart, and older generations were not that dumb.

In Just What Ways Are We Smarter?

On the other hand, we know that we must have gotten smarter because we know that school makes us smarter and we have had a lot more education than our forebears. So just how much smarter are we—and in what ways?

To help us answer this question, let's look at the scores on the WISC IQ test and on the most widely used "culture-free" test, namely, the Raven Progressive Matrices. Figure 3.1 shows the changes in scores over the period 1947–2002 for the Raven matrices; the full-scale WISC IQ (given to millions of children from six to sixteen years old); the five "performance" subtests of the

WISC that measure fluid intelligence (Picture Completion, Block Design, Object Assembly, Picture Arrangement, and Coding); two verbal subtests of the WISC that measure crystallized intelligence (Similarities and Comprehension); and three other WISC subtests that measure crystallized intelligence (Information, Vocabulary, and Arithmetic). Note that all scores on the tests and subtests are presented with the mean set equal to 100, to make comparisons among them easier.

The graph reveals a remarkable patchwork pattern. Scores on the Raven matrices and some subtests of the WISC have increased markedly while scores on others have hardly changed.

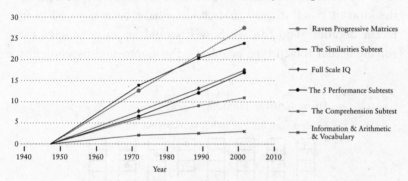

Figure 3.1. WISC full-scale IQ and WISC subtest scores and Raven Progressive Matrices scores from 1947 to 2002 for the United States. Reprinted by permission from Flynn (2007).

Let's first discuss the allegedly culture-free Raven matrices (see Figure 1.1). This test, which is supposed to reflect raw intelligence, that which is not susceptible to contamination by culture or school, shows an average gain in intelligence of more than 28 points! Grandchildren of the person in 1947 with an IQ of 100 now have an average IQ close to the official genius level as measured by that test. We can absolutely rule out the possibility that people's real intelligence, defined as general problem-solving ability, and so on, has increased by that much. We can also rule out the possibility that the Raven test is culture-free. It is absolutely drenched in culture. We know this because genes cannot have changed to such a significant degree during that period of time, nor have nutritional standards or any other biological factor that might affect intelligence.

So why has performance on the Raven test increased so much? We don't know, but we can engage in some informed guesswork. Developmental psychologist Clancy Blair and his colleagues have shown that the teaching of mathematics, beginning very early in elementary school and kindergarten, has shifted from instruction in mere counting and arithmetical operations to presentation of highly visual forms of objects and geometrical figures whose patterns children have to figure out. Figure 3.2 gives an example of the kind of visual display that children have been exposed to in recent decades. You can see how it would be an advantage for solving actual Raven-type problems. Developmental psychologist

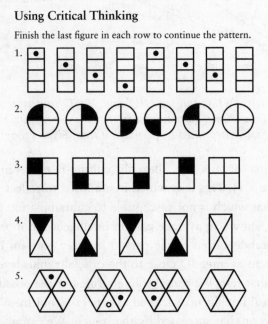

Using Critical Thinking

Finish the last figure in each row to continue the pattern.

Figure 3.2. Example of a visual problem used in a modern arithmetic textbook for very young children. Reprinted by permission from Eicholz (1991).

Wendy Williams has shown that classroom instruction in general emphasizes abstract perceptual tasks much more than in earlier eras. Blair and his coworkers have also shown that math problems are more likely than in the past to require multiple sequential operations of a sort that is essential to the working-memory capacity required to solve Raven problems.

But Figure 3.2 shows that it is not merely Raven-type problems but virtually the entire range of performance, fluid intelligence–type skills that have been improved by a culture that has been moving toward ever more visual forms of stimulation—textbooks, television, children's game books, and computers, including computer games. For example, all of the performance subtests of the WISC have a substantial visual component and many of them require multiple operations that have to be held in working memory for their solution.

We have every reason to believe that these sorts of visual exercises actually improve fluid-intelligence skills and the executive functions that underlie them, including working memory and control of attention. Researchers have shown, for example, that video-game players can attend to more things at once than can nonplayers. Video-game players can also ignore irrelevant stimuli more effectively than nonplayers and can see objects in a broader visual field than nonplayers. To make sure that what they have observed of computer-game players is not just a self-selection effect (with the fluid IQ hotshots being the ones most likely to play computer games in the first place), the researchers had nonplayers learn a game—Medal of Honor—that they thought would teach the attention-control skills of their video-game players, and had other nonplayers play a computer game that the researchers did not believe would be capable of teaching attention control, namely, Tetris. Subjects played the computer games for an hour a day for ten days. At the end of the period, Medal of Honor players did better on the attention-control tasks than did Tetris players.

Neuroscientists have shown that it is possible to use computer games to train very young children in executive functions of the kind underlying fluid intelligence. Researcher Rosario Rueda and her colleagues have focused on attention-control tasks. They had four-year-olds work on a variety of computer-based exercises over a period of five days. For example, the children had to keep a cartoon cat on grassy areas and out of muddy ones by using a joystick. The children also carried out anticipation exercises—e.g., learning how to forecast the movement of a duck across a pond—and they performed a task that required them to memorize attributes of different cartoon portraits. And the children had to pick out the larger of two arrays of numbers, with conflict being introduced by having the larger array made up of lower numbers. For example, the child had to pick nine number 4s as being a larger array than five number 7s. The children also had to complete an inhibitory-control exercise. They had to click a picture of a sheep as quickly as possible but withhold the click when the picture was of a wolf in sheep's clothing.

These attention-management and executive-control tasks had a very significant effect on children's performance on a matrices task of the kind used in the Raven test. The scores of trained children actually exceeded those for untrained children by more than .40 SD. Remarkably, there was a measurable effect on brain wave patterns. Electroencephalograms were recorded while the children performed some of the tasks. Activity in portions of the brain that mediate attention control was shifted so much that the patterns observed for the trained four-year-old children were more typical of six-year-old children than they were of untrained four-year-old children.

So we have every reason to believe that the culture is producing superior executive-control functions than were found for earlier eras, and that these altered executive functions are improving performance on fluid-intelligence tasks—certainly for Raven matrices and probably for other fluid-intelligence tasks such as those in the WISC performance package. As we might expect, by the way, computer-training tasks like those used by Rueda and colleagues can improve attention control for children with attention deficit hyperactivity disorder (ADHD). The ADHD researchers were also able to improve the working memory of adults without ADHD.

Cognitive neuroscientist Adele Diamond and her collegues have shown that mundane play activities can also increase executive functions. They structure play for nursery school children in ways that teach attention control and inhibition, crucial aspects of executive functioning. Children plan their play activities in an explicit way, learn to prompt themselves to act in certain ways with the use of memory aids, and get experience in turn-taking. These activities pay off in improving performance on standard executive-function measures.

So does the gain of 2 SDs for the Raven over fifty-five years and the gain of more than 1 SD for the WISC performance tests indicate truly massive shifts in intelligence? Probably not. They do indicate huge changes in skills that underlie some kinds of fluid intelligence, but they may not affect problem-solving skills in domains very far removed from those tasks. We don't know how broad a net is cast by these skills at this point.

It should be clear that any claims that the Raven matrices comprise a culture-free IQ test are now completely invalidated. Using that test to compare illiterate Amazon tribespeople or Africans with only a little Western-type schooling, with Americans or Swedes or Spaniards living in highly complex and educated and computerized cultures, is no longer intellectually supportable—if it ever was.

This is not to say, however, that smarter people within a culture do not do better on the Raven test than people who are not so smart. We know that smarter people do better now and that they did two generations ago. This is because scores on the Raven do and did predict academic competence and career success to a degree. It is just that the Raven and other fluid-intelligence tests are getting easier for everybody in the wake of cultural changes, which include the teaching of mathematics and computer activities.

Scores on two tests generally thought of as measuring verbal skills or crystallized intelligence have also improved a lot over time. Performance on the Similarities subtest of the WISC has increased by an amount equal to 24 IQ points over the fifty-five-year period from 1947 to 2002. To get a good score on the Similarities subtest, you have to be able to say that summer and winter are alike because they are both seasons—or if you are really clever, that they both have solstices. Only partial credit is given for saying that summer and winter are times of the year, and none at all is given for saying that "they are parts of nature" or that "they are windy." You also have to be able to say that revenge and forgiveness are alike because, for example, they represent choices about what to do if someone does something bad to you. You get only partial credit for saying that they are both decisions, or they are both actions that you take toward someone, and you get no credit at all for saying that they are feelings or types of resolutions. In short, you have to be able to abstract attributes from the test items and see some intersection of those attributes that is maximally interesting or informative.

Does the improvement in performance on the Similarities subtest mean that we have gotten 1.60 SDs smarter in the past two generations? No, but it does mean that we can think analytically in ways that help us to understand and generate metaphors and similes and that we have gained in our ability to categorize objects and events in ways that are relevant to scientific classifications. And these changes are of real significance.

Note that although the Similarities subtest is held to represent a form of crystallized intelligence, and answering the questions correctly does indeed rest on the storehouse of information that a person has, the subtest also has a significant fluid-intelligence component at the level of the more advanced items. You have memorized that summer and winter are seasons, but you have to infer on the spot just what features revenge and forgiveness have in common and assess which of those features are the most pertinent.

The other crystallized-intelligence subtest showing significant improvement in recent decades is the Comprehension test. In some ways I find this test the most convincing of all to build the case that people really have gotten smarter. It is impressive that young children today are more likely to understand why they should turn off electric lights when they are not using them, and it is impressive that somewhat older children are more likely to be able to say why people pay taxes. And the gains are quite significant—one-third of a standard deviation per thirty-year generation. To what should we attribute these gains? I don't know, but I suspect that television has a lot to do with it. Children can learn a lot about the way the world works by watching educational TV shows like *Sesame Street* and even some shows that are meant purely as entertainment.

The changes on the Comprehension subtest are not likely to be due to increased reading. For an explanation, let's turn to the three crystallized-intelligence tests, which have shown little change over time. The average score on the Vocabulary subtest has increased by a little more than .25 SD. This is not nothing, but it is not at all in line with the improvement on the Comprehension subtest. This seems understandable in view of the fact that people read less now than they did in earlier times. The percentage of seventeen-year-olds who read nothing at all for pleasure has doubled over the past twenty years. On the other hand, we do know from other evidence that younger children are learning to read well a little sooner than they did even thirty-five years ago. The National Assessment of Educational Progress (NAEP), which has tested nine-, thirteen-, and seventeen-year-olds every few years since the early 1970s, has established that nine-year-olds are gaining in reading proficiency at a rate of about .25 SD per generation. Thirteen-year-olds have gained a smidgen, and seventeen-year-olds nothing. In general, the reading results of the NAEP are pretty much in line with the Vocabulary subtest results. It is worth noting that the negligible improvement for older children is actually better than might be expected given that the level of textbooks for older children in the United States has been dumbed down by about two years in recent decades.

What is puzzling is how little improvement—none actually—has been achieved on the Information subtest in two generations. In some ways, this is not surprising in that children spend less time today memorizing facts. I was required to learn all the state capitals. Knowing that the capital of Kentucky is Frankfort and not Lexington has not proved particularly useful to me. On the other hand, it is a little puzzling that the average score on the Information subtest has gone up so little while that for Comprehension—understanding of why the world works as it does—went up so much.

Another conundrum is posed by the fact that the average score on the Arithmetic subtest of the WISC hasn't increased in the past thirty years, yet we have other evidence indicating that math skills have genuinely improved.

At the turn of the twentieth century, geometry was taught only to college students and upper-level high school students; in midcentury it was not generally taught before the tenth grade; now geometry is taught in the late junior high and early high school years. "Pre-geometry" perceptual material and calculation are taught in elementary school. It was understood in 1900 that calculus could not be taught before the senior year of college, and that was a time when fewer than 10 percent of Americans attended college. At midcentury it was taught in the early years at the better colleges and in the senior year at the best public and private high schools. Now calculus is routinely taught in the last year of high school, and at some elite schools it is taught as early as junior high. In 1920 fewer than 20 percent of Americans were educated through high school. By 1983, more than 80 percent were. So it seems very strange that math scores on the WISC have not improved in sixty years.

Yet there is substantial evidence that students are in fact better at math, at least prior to high school. The NAEP shows math scores improving by two-thirds of a standard deviation for nine-year-olds in the period 1978–2004. Scores for thirteen-year-olds improved by more than half a standard deviation. Scores for seventeen-year-olds improved by almost a quarter of a standard deviation. I suspect that the way to reconcile the WISC results with the NAEP results is to note that the WISC emphasizes rote application of learned arithmetic procedures, whereas the NAEP emphasizes mathematical reasoning and the sorts of multiple-operation problems that put a stress on working memory. But I hasten to say that I am far from confident about how to resolve the discrepancy.

Will IQ scores continue to go up without limit? Surely not, though there is no evidence of the increase slowing in this country. It does seem to have come to a halt in Scandinavia. In contrast, increases in IQ may have begun in the developing world. In a particular region of Kenya, seven-year-olds gained 1.70 SDs on the Raven Progressive Matrices test over a fourteen-year period and .50 SD on a test of verbal intelligence. Since it was seven-year-olds who were tested, only four months after they started school, it is unlikely that schooling played much of a role in the changes. Access to popular culture of the kind that has been increasing in developed nations, such as video games, accounts for little or none of the gain. Plausible explanations include increases in parental education, which were substantial during the period; improvements in nutrition, which were also marked; and reduced incidence of hookworm. A study on the Caribbean island of Dominica revealed an 18-point gain on the Raven Progressive Matrices and a 20-point gain on a vocabulary test over a twenty-five-year period.

What can we say about IQ gains then?

1. School definitely makes people smarter. The information and problem-solving skills learned in school result in a higher IQ score. A year of school is worth two years of age for IQ.
2. People's abilities to perform some of the tasks used to measure IQ have improved over time. This seems inevitable given that more people are getting educated, education is more and more geared toward the kinds of skills that lead to higher IQ scores, and some aspects of popular culture are intellectually challenging.
3. Some of the IQ gains (for example, on the Comprehension and Similarities subtests) clearly amount to real gains in intelligence for dealing with everyday-life problems.
4. Some of the gains in IQ are quite important in that they lead to improvements in academic achievement and should improve the ability to complete tasks involving abstraction, logic, and on-the-spot reasoning—the kind that are encountered in industry and science. Tests measuring such fluid-intelligence gains

include the Block Design, Object Assembly, Picture Arrangement, and Picture Completion subtests on the WISC, as well as
the Raven Progressive Matrices.

5. These fluid-intelligence gains probably don't contribute much
to the ability to perform everyday practical reasoning tasks.

6. The gains in IQ make it clear that performance, fluid intelligence–type tests such as the Raven matrices are not culture-free
measures of intelligence, a claim some IQ researchers still make.
Such fluid-intelligence tasks are much more culture-saturated
than are crystallized-intelligence tasks. In fact, the gains on
these tests raise a question as to whether a culture-free measure
of intelligence is possible.

7. Like the fact that schooling adds points to an IQ score, the fact
that gains have occurred over time in skills that society cares
about—both for everyday life and for advanced work in science,
industry, and other professions—establishes that people can
become smarter in very real and important ways.

8. Finally, the evidence speaks against two very pessimistic assertions made by Charles Murray. He has said that even a perfect
education is not going to make much difference for people in
the bottom half of the IQ distribution. But the average IQ of
people in the bottom half has improved by more than a standard deviation in sixty years, and their performance on the
Raven Progressive Matrices, the longtime gold standard of IQ,
has improved by more than 2 SDs. He has also said that because
higher-IQ people have fewer children than do lower-IQ people,
the average intelligence of the population must be going down.
The evidence suggests the opposite.

In the next chapter you will see if it is possible for schools to
do an even better job of making people smarter, and in Chapter 7
you will find out if it is possible to bring the IQ of people in the
bottom half of the IQ distribution closer to that for people in the
top half.

CHAPTER FOUR

Improving the Schools

I took a good deal o' pains with his eddication, sir; let him run in the streets when he was very young, and shift for hisself. It's the only way to make a boy sharp, sir.
—The elder Weller, Charles Dickens's
The Pickwick Papers, 1836

IF PEOPLE HAVE BECOME smarter over the last century, in part because they have had more and better schooling, can the schools be improved to produce an even bigger boost to intelligence? If so, what can be done to make them even better than they have been to this point? These questions are particularly pointed for Americans because the United States is well behind most of the developed world in its level of educational achievement. U.S. students who score in the 95th percentile on general knowledge assessments are comparable to students in the 75th percentile in some of the high-scoring countries. The most advanced U.S. students, the 5 percent taking advanced placement (AP) calculus and the 1 percent taking AP physics, score about as well as the top 10 to 20 percent of students in other countries. So, even a comparison of top performers reveals a substantial gap between the United States and other advanced countries.

There is plenty of room for improvement, so let's see what might be done to make kids smarter and more academically accomplished.

Does Money Matter?

There is some surprising evidence about what does *not* work. Since the 1970s, researchers have maintained that the amount of money spent on schools is not very closely related to their effectiveness. This conclusion is usually based on multiple-regression studies in which a large number of variables are put into the analysis to see which ones matter. The studies essentially ask if, net of everything else, the quantity of money spent makes a difference. That amounts to the following: If one is looking at schools where the population consists of rich majority kids in the Northeast, for example, differences in amount of money spent do not matter much, and if one is looking at schools where the population consists of poor minority kids in the South, differences in amount of money spent do not matter much.

It seems obvious that the sheer amount of money spent would have no effect by itself on education. Judges sometimes force communities to spend as much for schools attended by poor students as for schools attended by better-off students. Such additional funds may be spent without much careful thought and planning, and when this happens there is little evidence that the scores for the poor improve very much. The classic case of this occurred in Kansas City, where judicial orders increased hugely the amount of money pumped into the schools. Olympic-size pools were built, state-of-the-art science labs were provided, and a computer was given to every student. The result: no improvement in scores. Money by itself does nothing to improve student performance, especially when administrators are incompetent and corrupt, as is the case in some big-city districts. Other evidence about the effect of money comes from a lack of association between the amount of money spent per student in developed nations and achievement scores on tests such as the TIMSS (Trends in International Mathematics and Science Study). Some high-scoring nations are below the median on per capita expenditures, and some low-scoring nations are above the median on per capita expenditures.

But this does not mean that money never matters. In a later chapter you will find out whether there is any evidence of ways to improve education, especially for the poor and for minorities, by spending more money than we now do.

Vouchers and Charters

Many school critics have urged a voucher system—giving parents money so they can send their kids to private schools. But researchers should not compare children whose parents are offered vouchers and actually take them with children who were not offered vouchers, because of the problem of self-selection: it might be that the more educated, intelligent, and motivated parents make use of the vouchers they were offered. The authors who claim that voucher programs have big effects did indeed compare children whose families were offered the vouchers and actually made use of them, with children who were not offered the vouchers. Such studies found as much as a one-third reduction in the black/white gap in test scores. But at least some of this "gain" really only represents a difference between children whose parents cared enough to carry through on enrollment and children whose parents might or might not have carried through if given the chance. When researchers maintain that they can control for this effect by procedures such as matching the voucher-users with students whose families were never offered the vouchers but who are comparable on a variety of dimensions, they are talking through their hat. When they let self-selection enter the front door, they cannot get rid of the problem by matching through the back door. In other words, parents who take advantage of an offer that sounds good for their children may be different from otherwise very similar parents in a control group who might or might not have taken advantage of that offer.

When studies on the effects of vouchers are done properly—comparing the achievement scores of all children who were offered the vouchers (most of whom accept them) with those of similar students who entered into a lottery for the vouchers but lost and so were not offered the vouchers—the evidence indicates that vouchers result in perhaps .10 SD better performance. (Such a comparison is called an intent-to-treat design: students in the treatment group are included in the analysis regardless of whether they accepted the treatment. This almost inevitably underestimates the treatment effect if there is one, but avoids the deadly self-selection that occurs when the only students compared to controls are those whose parents cared enough to see that they took advantage of the treatment). The disappointing results of this study of course do not prove that private schools are not capable of doing a better job than public schools, just that there is not much convincing evidence that the private schools studied to date do a better job.

There is also little evidence that charter schools in general are particularly beneficial. These schools are publicly funded but have been freed from some of the rules and laws that govern other public schools. In exchange, such schools set out in their charter some assurances of accountability for producing certain results. Reasonable experiments, with random assignment, have compared charter schools with other public schools. Unfortunately, charter schools seem to be only slightly better than the public schools that the children would otherwise have attended—at least for the first few years of their operation. For math and reading, the improvement on standardized tests runs just a few percentile points higher for young students in a charter school. Older students entering a charter school for the first time in later grades actually do worse than the public school students. There is, however, some evidence that after charter schools have been in operation for ten years or so, they may have as much as a 10 percent edge on public schools. None of this should be taken to mean that charter schools cannot do a better job than the public schools—just that not many have to this point. In a later chapter you will see that there is at least one very happy exception to this generalization.

Class Size

How about class size? Are smaller classes better? Here we have a real conflict of evidence. The multiple-regression analysts tend to tell us that class size makes little difference to student achievement. On the other hand, economist Alan Krueger, the researcher who has written the most about class-size effects, maintains that 70 percent of studies that looked at class size find a positive effect, and the better the journal in which the study was published, the more likely it was to show a positive effect. And there is one study that, in Krueger's opinion and in mine, is more relevant than all the others put together. For this experiment, which was conducted in Tennessee in the 1980s, teachers and students in kindergarten through third grade were randomly assigned to classes of regular size (twenty-two pupils on average) or to smaller classes (fifteen pupils on average). The children in the smaller classes did better on standardized achievement tests, the improvement varying on average from .19 to .28 SD depending on subject and class size. In other words, being in a smaller class was typically enough to shift a child on average from the 50th percentile for achievement to a little less than the 60th percentile. The effects persisted to a significant degree through at least the seventh grade. And the effects of a smaller class were bigger for poor and minority children than for middle-class and white children.

Teachers Matter

How about teachers? Do they matter? Surely some teachers are better than others? Indeed so, but the fact that a teacher is certified is not proof that the teacher is particularly good at the job. Nor, surprisingly, is possession of a master's degree.

Nevertheless, there is plenty of evidence that teachers can make a difference. First, experience matters. The average difference in reading achievement scores for children taught by teachers with one year versus ten years of experience is .17 SD—an expected benefit for experience of about 7 percentile points on achievement tests. But note that most of the difference between more and less experience occurs in the first year of teaching. So it is definitely worth trying to avoid having your child put in a class with a rookie teacher.

It is possible to measure teacher quality by defining it as the value a teacher adds to the achievement scores of the average student in his or her class over and above the achievement scores for the average student the previous year. A teacher produces high added value to the extent that students do better than expected given how well the students performed relative to their peers the previous year or years. Defined in this way, 1 SD in teacher quality is associated with approximately .20 SD in achievement scores. But this is an estimate of the difference teachers make within a given school. Since we know that teachers in some schools are better on average than teachers in other schools, we can be sure that the within-school difference is a minimal estimate of the difference made by teacher quality. Economist Eric Hanushek's estimate of the impact of teacher quality, across teachers and across schools, is .27 SD—more than enough to raise a child from the 50th percentile to the 60th percentile in achievement. The difference in achievement scores that could be expected for a child who had teachers rated 1 SD or more above the mean for quality throughout elementary school and a child who had teachers rated 1 SD or more below the mean for quality is hard to calculate but clearly could be very great.

But statistics like these on the quality of teachers pale in their implications beside stories that people tell about the importance of individual teachers to their lives. Most people I know believe that at least one or two of their teachers made a great difference to them, and there is no reason to doubt them. A study on the effects of one particular first-grade teacher gives substance to the anecdotes.

The teacher, Miss A, taught for thirty-four years in a school serving a low-socioeconomic-status population, a third of whom were black. Sixty students who had been at the school over a period of eleven years were interviewed when they were adults. A third of them had had Miss A for a teacher in first grade. Among those who had had teachers other than Miss A, 31 percent could not remember the teacher's name. All of the students who had had Miss A could remember her name. Whether or not they could remember their teacher's name, only a third of students from the other classes rated their teacher as having been very good or excellent. Three-quarters of Miss A's students gave her a rating that high. Twenty-five percent of children with another teacher gave the teacher an A for effort; 71 percent of Miss A's students did. Asked by a researcher how Miss A taught, a colleague answered, "With a lot of love." The colleague also reported that Miss A expressed confidence that all her students could learn—no child was going to leave her classroom without being able to read. She stayed after school to help slow learners. She shared her lunch with children who had forgotten theirs.

Outcomes in elementary school, in youth, and in adult life were much better for Miss A's students. Two-thirds of her former students scored in the top third in achievement in second grade, compared to only 28 percent of students of other teachers. The students' status as adults was also measured, by such things as grade of education completed, occupational attainment, and condition of the home. Of Miss A's students, 64 percent were at the highest level of status; of other teachers' students, only 29 percent were at the highest level.

This story, inasmuch as it is based on a single individual, could be discounted if there were not other, more convincing, quantitative evidence about the importance of teachers, especially the first-grade teacher. But in fact such evidence exists. Education researchers Bridget Hamre and Robert Pianta had access to the huge longitudinal Study of Early Child Care sponsored by the National Institute of Child Health and Human Development. They tracked about nine hundred children in their progression from kindergarten through the end of first grade. Some of the children were considered to be at risk of poor adjustment to the school situation on the basis of socioeconomic status as indicated by the mother's educational attainment, and some were considered to be at risk based on the kindergarten teacher's reports of problems displayed in behavior, attention, and academic performance.

The quality of the first-grade experience of each of the children was evaluated by observers sitting in on each class for three hours. The classrooms differed along two related, but somewhat statistically independent, dimensions. The first dimension was labeled "instructional support" and consisted of the sum of ratings of quality of literacy instruction, quality of evaluative feedback to the children, degree to which conversation was concerned with instruction, and encouragement of child responsibility. The second dimension was labeled "emotional support" and consisted of the sum of ratings of emotional climate, effective management of the classroom, involved as opposed to detached relationships with the children, and relative absence of intrusiveness.

Classrooms were divided into three levels on the basis of quality of instructional support. All children were tested on a widely used measure of ability—the Woodcock-Johnson tests of cognitive ability and academic achievement.

When a child who was at risk on the basis of the mother's relatively low educational attainment was placed in a classroom providing low instructional support, the child's achievement score at the end of first grade was more than .40 SD lower than would be expected if the child had been placed in one of the classes having relatively high instructional support. When a child of similar risk was placed in one of the classrooms that provided high instructional support, the child actually functioned at the same level on average as children of highly educated parents in those classrooms.

Classrooms were also divided into three levels on the basis of quality of emotional support. (In general, classrooms were more likely to be at a high level of emotional support if they were rated high on instructional support and at a low level of emotional support if they were rated low on instructional support, but the overlap was by no means complete.) Children who were at risk of failing on the basis of poor social and emotional functioning in kindergarten, and who were placed in a classroom having either low or moderate emotional support, scored about .40 SD worse than would be expected if they had been placed in a classroom having high emotional support. Children who were at risk on the basis of kindergarten functioning and placed in a highly supportive classroom did just as well as children who were not at risk and were in a highly supportive classroom.

The latter finding may be even more important than it sounds. Hamre and Pianta found, in an earlier study, that relationship problems in kindergarten are associated with academic problems throughout school. The findings about teacher quality in the first grade suggest that children with such problems may have a downward path deflected upward very early if they are placed in the right first-grade classroom.

Principals know about the quality of teachers in their schools— at least at the extremes. But there is little evidence that principals seek out or reward high-quality teachers. In fact, for teachers in the public schools it would be difficult for them to do so. Union rules tend to enforce similar pay standards for teachers and tend to allow for differences having to do only with seniority, certification, and the possession of higher degrees. There is little evidence, as I have noted, that certification and higher degrees are associated with better teaching, and beyond the first year or so of teaching, neither is seniority. Researchers who are aware of these facts tend to fall into one of two policy camps: (1) change the rules and make it possible to reward teachers for good teaching or (2) concede that this is difficult to accomplish politically and instead focus on teaching teachers how to teach better.

Everyone can agree that one of the most important things we can do for education is to improve teaching. Perhaps one place to start would be in schools of education. A common complaint from new teachers is that they received too much theory in education classes and too little experience or practical training. Another route to improving teaching would be to provide incentives for good teaching. Israeli researchers conducted a study of two different incentive programs. One program awarded bonuses to teachers at schools in the top third of performance based on their students' achievement. The bonuses amounted to 1 to 3 percent of base salary. The other program gave greater resources, mainly teacher-training programs and reduced teaching time, to the winning schools. Both programs resulted in improvements in student scores and in dropout rates, but the salary incentive program was the more cost-effective. The study indicates that providing incentives at the school level might avoid union problems. It is conceivable that if such incentives were perceived as add-ons, and winners were at the school level rather than the individual level, teachers and their unions might accept the competition with its reward structure of "possible gain vs. no loss." Just how to assess school quality, however, could be as much a political football as assessing individual teacher quality.

"Effective Schools"

Until relatively recently, there was little research showing convincingly that some educational techniques are more effective than others. There has long been a huge literature on what are called "effective schools"—schools that go beyond expectations for student achievement. But this evidence rarely rises above the anecdotal. It says that good schools are characterized by principals who convey a conviction that most children are capable of learning; who carefully pick their teachers and carefully monitor them; who find ways to encourage failing teachers to leave the school; who emphasize curriculum and instructional strategies; who monitor student performance data to see if the instructional strategies are working; and who seek parents' involvement in their children's education. Teachers in effective schools are less isolated and are more likely to talk shop; they are more likely to be evaluated and to appreciate being evaluated; and they monitor student performance data to see if the instructional strategies are working.

Much of the literature on effective schools deals with schools serving disadvantaged populations. The usual claim is that schools that emphasize the basics more are more effective with these populations. On the other hand, some people claim that an enriched curriculum more characteristic of good private schools can be highly effective.

In short, schools with better outcomes have better principals with better strategies and teachers who are more committed to seeing that their students flourish. But I find little in the literature to convince me that these things cause successful outcomes rather than merely reflecting the fact that the population being served is one that is easier to help. Principals are going to look good if their students are easy to work with. If students are sufficiently troublesome, focusing on curriculum and teacher evaluation may be secondary to enforcing discipline. So while there are many stories about remarkable schools, there are not many hard facts that leave me confident about how to improve underperforming schools.

Educational Research and Its Enemies

Despite the hundreds of millions of dollars spent on innovative educational programs, and the hundreds upon hundreds of studies evaluating them, the situation in educational research is scandalous. Research is mostly anecdotal, and most self-styled evaluators of educational programs are actually opposed to the experimental method, that is, providing one educational technique to children randomly selected from some population and providing a comparison technique to other randomly selected children. Very little research rises to the level of being scientifically acceptable.

The situation is as shocking as it would be if pharmaceutical companies were to routinely peddle their medicines without having them backed by evaluation research that went beyond haphazardly giving the medicine to some individuals with a given illness and reporting the percentage of patients who got better (without knowledge of the percentage who would have gotten better without any treatment at all). Only drug trials that identify a patient population and then randomly assign some patients to the treatment condition and some to the no-treatment or alternative-treatment condition count as adequate research. Yet this standard is almost never met in research on educational interventions. Anyone who tells you that studies lacking random assignment are just as good or better for evaluating educational research should be asked why educational research should be held to different standards than drug research.

Plenty of reasons are given for not using the experimental method. There is the claim that ethical considerations demand that the most needy receive the treatment. But ethical considerations require that the investigators establish that they are doing good— which can be done by finding the most needy and assigning half of them to the treatment group and half to the control group. The most specious reason given is that experiments somehow block finding out just what elements of a treatment cause it to be effective. Without knowing *that* a treatment is effective, there can be no way of knowing *how* it is effective.

Recent research on schools has employed at least some form of control. In some studies, investigators get schools to agree to accept an intervention, for example, a new type of computer instruction for math, and then compare performance at those schools with that at schools that are similar on a predetermined set of criteria, such as the social class and race of students, but that were not offered the intervention. This type of research is better than nothing, but not by much. It is susceptible to the self-selection problem: the schools that are offered the intervention may be systematically different in some unknown ways from those not offered the intervention. The problem is particularly acute when there is literal self-selection, that is, when only some of the schools offered the intervention accept it. The schools that accept the intervention may rate better on some of the relevant dimensions than those that do not accept it.

Also inadequate are studies that simply report scores at a school before the intervention began and compare them with scores after the intervention began. These studies generally yield effect sizes that are substantially greater than those found by studies comparing the schools that had the intervention with presumably comparable schools that did not. An exception to this rule exists when gains after an intervention are extremely large—and discontinuous with what could have been expected if there had been no intervention. Under such circumstances a claim of effectiveness can sometimes be persuasive.

Whole-School Interventions

Some of the programs that have been evaluated are whole-school interventions, known as comprehensive school reform. Educational psychologist Geoffrey Borman and his colleagues reviewed a number of the most promising of these programs, and my account of them rests on their review. I report on only the programs that have had three or more tests by independent third parties, and those in which schools having the intervention were compared to control schools. Virtually none of the comparisons are based on a random assignment of schools or students.

One whole-school program—Success for All—has been evaluated twenty-five times in comparison studies by third-party investigators. Hundreds of schools now participate in this program, which is managed by a private foundation. Success for All equips schools with its own curriculum materials, including teacher manuals. It offers a great deal of teacher training in reading, writing, and language arts, as well as twenty-six days of on-site professional development. The program emphasizes assessment of student outcomes and school organization and provides facilitators for each school. One-on-one tutoring is given to students having problems in reading. Outreach to parents is emphasized. Originally the program was available for pre-kindergarten through sixth grade, but it now has a junior high school component as well. Some schools also participate in a broader version of Success for All called Roots & Wings, which offers programs in math, science, and social studies.

Success for All seems to make a difference in achievement outcomes, but the better designed and more independent the evaluation, the weaker the effects appear to be—an average effect size
of only .08 SD was found across all investigations. A particularly
well-designed, fully randomized study, however, revealed an average effect size of .27 SD for reading in children assigned to the
program from kindergarten through second grade. Moreover,
four of the evaluations included the broader Roots & Wings
program, and these evaluations produced an eye-popping average
improvement of .77 SD. A couple of highly convincing, independent, random-assignment studies would be required to make an
effect this large seem very plausible. Despite all the research, the
final word on Success for All is not yet clear. And it will not be
until more solid, random-assignment research is done.

Another very well-known program, the School Development
Program, was founded more than thirty years ago by James
Comer, a psychiatrist at Yale University. This program does not
specify particular curricula or instructional methods but tries to
build good relations among school staff, parents, and communities and sets up health interventions for students. It provides teams
that develop and carry out specific reforms related to the assessed
needs of individual schools. The third-party comparison studies
show an effect of only .11 SD.

Slightly more successful is the Direct Instruction intervention
for elementary schools, primarily those serving a disadvantaged
clientele. Direct Instruction reading and math programs are distributed by McGraw-Hill publishing company, and some teacher
training is offered. But the full program requires contracting with
providers of extensive professional curriculum development and
teacher training. The lesson plans are highly scripted and require
extensive writing. Lessons are carried out in small groups organized by performance level, and student progress is frequently
assessed. The third-party comparison studies show an effect size
of .15 SD.

The cost of some of the whole-school interventions can be

rather high. So a thorough cost-benefit evaluation is called for in addition to attending to established effect sizes. In fairness to these programs, and to many others that have been less extensively evaluated, some of the assessments are of interventions that have not been well implemented at some schools in the sample. One can always get null results for an intervention if it is handled badly enough, and the good, the bad, and the ugly all get averaged together to get effect-size estimates.

Instructional Techniques

A number of studies of specific instructional technologies have also been carried out. Evaluation researcher James Kulik reviewed a large number of so-called integrated learning systems. These computer software systems give students course materials customized to their level of attainment, keep records of how the students are doing, and provide substantial feedback on performance. Kulik reached some very clear conclusions. Across sixteen well-controlled studies of mathematics programs, he found computerized instruction to have a median effect size of .40 SD. This very large effect is educationally significant, and the costs of the program after computer purchase are low. Word-processing programs, by teaching writing, also have a significant effect on learning how to read. The effect may be as much as .25 SD in the upper grades and much larger in kindergarten and the first grade. (Computerized reading programs that attempt to teach reading without an emphasis on writing have a median effect size of only .06 SD.) Finally, computer tutoring is apparently effective for teaching the natural and social sciences. The median effect size for those fields was .59 SD, a very significant value. Computer tutoring also resulted in much more favorable attitudes toward the particular science being taught: the effect size was 1.10 SD.

Some of the most impressive intervention work to date goes by the name of "cooperative learning." This refers to classroom techniques where students work together in small groups, helping each other to learn a body of material. The technique can be applied to any subject matter. Students can reach the group's educational goal only if all of them bring their necessary part to the enterprise. It can be used for grades 2 through 12. Education researcher Robert Slavin wrote an entire book on these techniques. In order to be included in Slavin's survey, a study had to have a control group learning the same material, and there had to be an adequate control, either in the form of randomization at the individual, class, or school level, or by means of matching— that is, finding groups of students who were similar on many criteria to those who were offered the program. As it happens, results using the far superior randomization methods were approximately the same as results using the matching method. In one variant, Student Teams Achievement Divisions, students are assigned to four-member teams (which are usually heterogeneous with respect to prior achievement level, ethnicity, or both). The students work together studying the materials and are then assessed individually. Slavin reported studies indicating that this technique has an effect size of more than .30 SD on standardized tests. One particularly impressive technique is the "structured dyad" method: one student is the tutor and one is the tutored; roles are then switched. There are a variety of ways to accomplish this, but reported effect sizes run very high for all of them. Enough studies of this general technique have been conducted to indicate that instruction in grades 2 through 12 ought to include a cooperative learning component.

Summing Up the Research

So what do we have? Can schools apply new procedures and improve their ability to make people smarter? There are lots of answers of the "no" variety, or at least of the "not yet" or "not much" variety. Money per se does not make a great deal of difference. Vouchers and charter schools have not produced substantially better academic achievement than regular public schools.

Teacher certification and credentialization are unrelated to student outcomes. Teacher experience does count—at least up to a point: first-year teachers are not as good as they will be, and learning to teach continues for a few years.

Teacher quality matters a lot: some teachers are just a lot better than others. But the current systems are not good at rewarding the best teachers or weeding out the worst ones. The research is still young, but there is at least some evidence that providing incentives to teachers at schools that succeed the most results in improved educational outcomes, and it is possible to imagine schemes to give incentives that might not encounter serious political problems.

The "effective schools" literature tells us what the principals and teachers at the schools with better outcomes are like, but it does not tell us about causality: To what extent do dedicated principals and focused teachers make for better schools? And to what extent is it the more teachable clientele that makes it possible for principals to look focused and teachers to look dedicated? Some whole-school interventions have been effective, but there is little evidence to date that any of these produce very big effects.

Extremely promising evidence shows computerized instruction to be very effective, especially for mathematics and science training. Cooperative learning, in which students work together toward common educational goals, also is very promising.

In an extremely welcome development, the U.S. Department of Education now runs what is called the What Works Clearinghouse. This service identifies interventions that evaluation studies have shown to be effective. Unfortunately, the designs of these studies typically fall short of randomized-assignment experiments but at least they are far above the level of anecdotal reports. All of the research conducted for the clearinghouse is supposed to be at least "quasi-experimental studies of especially strong design." These standards are well above most of what passes for evaluation research. We can hope that, eventually, educators will undertake only interventions that have been declared effective by the What Works Clearinghouse or will have to answer to an outraged public.

Enriching the Curriculum: Effects on Skills and IQ

But what would happen if we really pulled out all the stops and tried to teach kids general problem-solving skills that could make them genuinely smarter than they would be by studying materials that are typical for their grade level? A tantalizing answer to this question comes in the form of a massively ambitious study in Venezuela headed by, of all people, Richard Herrnstein, coauthor of the highly pessimistic *Bell Curve* book. Herrnstein and his coworkers devised a very advanced set of materials geared to teaching seventh-graders fundamental concepts of problem solving that were not targeted to any particular subject matter. In effect, they tried to make the children smarter by giving them handy implements for their intellectual tool kits.

The concepts and procedures taught were closer to high school or college level than to junior high level. The researchers gave sixty different 45-minute lessons on topics such as discovering the basics of classification and hypothesis testing, discovering properties of dimensions that can be ordered in some way, analyzing analogies, exploring the structure of simple propositions, understanding the principles of logic, constructing and evaluating complex arguments, learning how to trade off the desirability of outcomes against their probability, and evaluating the credibility and relevance of data. These tools are usually the side effects of learning about a subject or discipline rather than something teachers try to teach explicitly. Can we teach the tools directly—even to children—and show that they generalize to new problems with content different from the problems used to teach them?

In a word, yes. The instruction resulted in big changes in children's ability to solve problems that the new skills were designed to improve. Some of the effect sizes were as follows: for language comprehension, .62 SD; for learning how to represent "problem spaces," .46 SD; for decision making, .77 SD; and for inventive thinking, .50 SD. In short, general problem-solving skills can be taught, and taught moreover in a brief period of time.

How about "real" intelligence, as measured by IQ tests? Can teaching problem-solving skills improve IQ? I do not recognize IQ tests as being more than just one particular way to measure intelligence, rather than *the* way. If we can improve people's reasoning and decision making, I don't care whether this makes them perform better on IQ tests or not. But in fact, on a representative set of general-abilities tests, the score for the experimental group in the Venezuelan study increased .35 SD on average as compared to a control group. On a typical IQ test called the Otis-Lennon School Ability Test, IQ in the experimental group gained an average of .43 SD compared with the IQ in the control group. Even on a highly specialized, largely spatial IQ test similar to the Raven Progressive Matrices called the Cattell Culture Fair Intelligence Test, the gain was .11 SD. In short, whether intelligence is measured by general problem-solving skills of the sort that Herrnstein and his colleagues taught or by traditional IQ tests, the training had a very big effect.

You will want to know where the investigators went from there. Did they develop a yet more sophisticated tool box for eighth-graders? Unfortunately, the government in Venezuela changed, and increasing the intelligence of junior high school students was no longer a priority. I must say, however, that in light of the spectacular success of the program, it is amazing and dismaying to me that no one picked up where the Venezuela project left off.

Effective Tutoring

Finally, let's remember that a lot of teaching is in the form of one-on-one tutoring. And not surprisingly, tutors differ a lot in their effectiveness. In fact, Mark Lepper and his colleagues found that college student tutors and other tutors of elementary school students range from virtually ineffectual to extremely helpful. And they discovered some intriguing characteristics of effective tutors.

First, how do you go about being an *in*effective tutor? A sure-fire way to do it is to regard yourself as a debugger. You explicitly tell the student she has made a mistake and you give direct guidance in how to fix the error, preferably by using an abstractly stated rule. None of Lepper's most effective tutors took such a strictly cognitive, error-correction stance.

How do you go about being an effective tutor? Lepper gave us five Cs.

You foster a sense of *control* in the student, making the student feel that she has command of the material.

You *challenge* the student—but at a level of difficulty that is within the student's capability.

You instill *confidence* in the student, by maximizing success (expressing confidence in the student, assuring the student that the problem she just solved was a difficult one) and by minimizing failure (providing excuses for mistakes and emphasizing the part of the problem the student got right).

You foster *curiosity* by using Socratic methods (asking leading questions) and by linking the problem to other problems the student has seen that appear on the surface to be different.

You *contextualize* by placing the problem in a real-world context or in a context from a movie or TV show.

Expert tutors have a number of strategies that set them apart. They do not bother to correct minor errors like forgetting to put down a "plus" sign. They try to head the student off at the pass when she is about to make a mistake and attempt to prevent it from occurring. Or sometimes they let the student make the mistake when they think it can provide a valuable learning experience. They never dumb down the material for the sake of self-esteem, but instead change the way they present it. Most of what expert tutors do is ask questions. They ask leading questions. They ask students to explain their reasoning. They are actually less likely to give *positive* feedback than are less effective tutors, because, Lepper theorizes, this makes the tutoring session feel too evaluative. And finally, expert tutors are always nurturing and empathetic.

All of this could be taught to aspiring tutors—and to aspiring teachers in the schools.

In sum: We know that schools can do a lot better job at education than most are doing. We know a lot about what is effective and what is not. There can be no excuses for schools' failure to guide instruction in the light of what research shows to be effective.

But instruction is more effective for some clienteles than for others. Let's look at two clienteles in the next two chapters and then consider what can be done to improve the situation in the chapter following those.

CHAPTER FIVE

Social Class and Cognitive Culture

> *... the class structure of modern society is essentially a function of the innately differing intellectual and other qualities of the people making up these classes ...*
>
> —H. J. Eysenck, *The Inequality of Man* (1973)

THE VIEWS OF H. J. EYSENCK about class and intelligence are representative of much of mainstream opinion among IQ experts. Social class is a consequence of intelligence. The poor are poor because they are not intelligent, and neither money, nor class, nor parenting practices play much of a role in making some people more intelligent than others. Class is mostly a matter of genes.

In the chapter on heredity and mutability, we saw how utterly mistaken these views are. Undoubtedly, people of different social classes have different genotypes for intelligence on average, but there is an enormous causal influence of class on intelligence. As we saw in Chapter 2 on heritability, a lower-class child who grows up in an upper-middle-class family has an IQ 12 to 18 points higher on average than a lower-class child who grows up in a lower-class family. We can estimate that the lower-class families in these studies were at the bottom 15 percent or so of the social-class ladder, and the upper-middle-class families at the top 15 percent or so. The average IQ for the children of people in the lowest third of the socioeconomic totem pole is about 95, and the average for the children of those in the highest third is about 105. That 10-point difference is the result of all factors operating to push the social classes apart on IQ: genes; prenatal, perinatal, and postnatal biological factors; and all social factors associated with class, including quality of neighborhoods and schools and parenting practices. The results of the adoption studies indicate that postnatal environmental factors—biological and social ones combined—probably outweigh the genetic factors.

Eysenck and the other hereditarians could hardly be more wrong. Being poor versus well-off has a huge impact on intelligence.

In this chapter I highlight some of the differences in social class that directly influence intelligence. I will not be able to quantify exactly how much difference a particular factor makes. But we know that each of a very large number of factors has at least some influence on intelligence. Moreover, we know that some of these factors would be ameliorated if the poor were better off financially. And we know that there is a lot of room for improvement in the plight of the poor and the working class in the United States. Their economic situation is substantially worse than that in most developed nations.

We also know that parenting is different across the social classes. People of lower socioeconomic status (SES) are tacitly preparing their children for different occupational and social roles than are people of higher SES. Educators need to know the ways in which poorer children are ill-prepared for school in order to be able to help them achieve academically. Without improvement in cognitive functioning, countless numbers of poor people will be unable to take advantage of the jobs that exist in the new information economy.

First, some definitions. *Poor* refers to the habitually unemployed, people chronically on welfare, and nonskilled workers. *Working class* refers to skilled and semi-skilled workers such as mechanics and workers in service occupations and lower-level clerical jobs. I refer to these two classes as one group, lower-SES people. *Middle class* refers to higher-level clerical jobs, teaching jobs, and supervisory and lower-level managerial positions. *Upper-middle class* refers to professional and higher-level managerial jobs. I refer to these two classes as a group, higher-SES people. These definitions are somewhat arbitrary; there are other ways of carving up the domain of social class, but these definitions map onto the way the social classes were defined in studies referred to in Chapter 2.

In this chapter the differences I present between lower- and higher-SES people hold regardless of race. In the next chapter I discuss how race and class interact.

Environmental Factors of a Biological Nature

Being poor is associated with many environmental factors of a biological nature that lower IQ and academic achievement.

Some of the differences between people of lower and higher SES may have to do with nutrition. Although the available evidence suggests that poor nutrition in the mother prenatally has no effect on intelligence, when food supplements are given to children who are living in hunger, they show gains in IQ. It is not clear that nutrition differences between the social classes in the West are sufficient to contribute significantly to differences in IQ between the classes, but there is some hunger among a small percentage of the poor. Even if hunger is rare, we know that lower-SES children are more likely to have vitamin and mineral deficits. And there is evidence that supplements of vitamins and minerals improve the IQs of children who are lacking in them.

The effects of lead are very deleterious for IQ, and inner-city-dwelling kids are exposed to more lead in the form of pollution and old peeling paint than are middle-class and suburban children.

Even use of less than two ounces of alcohol by pregnant women has a negative effect on IQ. Children whose mothers drank during pregnancy have more difficulty in school because of attention and memory difficulties and poorer ability to reason. Lower-SES women are more likely to drink to excess during pregnancy than are higher-SES women.

Lower-SES people have poorer health, and poor health is an impediment to learning in many ways. A sick child will have a harder time learning than a healthy one. Specific health problems that are more common among the poor, and associated with low IQ and academic performance, include poor dental health, more exposure to smoke and pollution with consequent susceptibility to asthma, poorer vision, and poorer hearing.

Low birth weight is more common for lower-SES babies, and low birth weight is associated with lower IQ. Some pesticides that are being phased out, but are still not uncommon in lower-SES households, are associated with smaller head circumference and lower IQ.

One biological factor that may be important is breast-feeding. Lower-SES mothers are less likely to breast-feed. For children with the most common genetic makeups, breast-feeding seems to add about 6 IQ points over and above what the IQ would be without breast-feeding. Breast-feeding may exert its beneficial effect on brain development through the actions of particular fatty acids that are found in human breast milk but not in cow's milk and not in formula. The claim that the relationship between breast-feeding and IQ is causal is still in dispute. One study finds that children who were breast-fed have no higher IQ than their siblings who were not breast-fed. If the relationship is in fact causal, though, it could account for as much as 2 points in the IQ difference between higher- and lower-SES individuals.

Medical care is worse for the poor, and this not only compounds the problems associated with poor hearing and vision and asthma, as well as all the other biological factors, but also creates additional problems of its own. Lower-SES people are almost twice as likely to have no medical insurance. Even if they had insurance, working-class parents are less likely to be able to take their children to the doctor because they would lose wages or be absent from work, which would be a cause for discipline. In addition, doctors are much less available for the poor than for the middle class; there are three times as many doctors working around white, nonpoor neighborhoods than around poor, non-white neighborhoods.

Some of the environmental factors affecting biological processes are severely damaging but uncommon even among lower-SES people, for example, lead poisoning and fetal exposure to alcohol. Some are moderately damaging and are actually fairly common among lower-SES people, for example, exposure to polluted air, which can cause asthma. We do not know exactly how much each of these factors contributes to the IQ and achievement gaps. However, we cannot simply add up the differences made by each one and assume that they account for a very large fraction of the IQ difference between low-SES and high-SES classes. This is because the deficits are undoubtedly correlated with one another—children born to alcoholic mothers are also more likely to be exposed to peeling paint and to have poorer dental health. The deficits can't be regarded as additive.

Environmental Factors of a Social Nature

Other environmental factors are not biological in the first instance but undoubtedly have seriously damaging effects, possibly mediated by brain physiology.

One such harmful circumstance is that lower-SES children move from one house to another much more frequently than higher-SES children. As a result, they have to deal with the stress of moving more often, and they are put into classroom situations for which they are unprepared or are forced to repeat material that they already know. Even when the child stays put, other children are rotating in and out of the classroom and make the environment unstable and teaching more difficult.

Lower-SES children are more likely to have behavior problems, which are disruptive to one degree or another for all who have to deal with such children. Instability of all kinds is more common for lower-SES children than for higher-SES children. Lower-class neighborhoods are in general more stressful, and lower-class homes are more susceptible to turmoil and strife.

Compared with higher-SES parents, lower-SES parents are less likely to be warm and supportive of their children and are more likely to punish infractions harshly. Developmental psychologist Vonnie McLoyd showed that lower-SES parents are more likely to raise their children in punitive and stressful ways than is characteristic of higher-SES parents.

Early emotional trauma damages the prefrontal cortex, which (you may recall from Chapter 1) is heavily implicated in fluid intelligence. We do not know exactly how much stress is necessary to damage the central nervous system, but it is possible that the higher stress associated with lower-SES parenting, together with the other stresses of lower-class life, does produce such damage, at least at the extreme. Fluid intelligence is particularly important to learning and to school achievement in the early grades.

Of course, not all lower-SES children have such extreme difficulties. Undoubtedly most have loving families who deal with them in kindly fashion and are deeply concerned with their physical and intellectual development. Many lower-SES children live in neighborhoods that pose few terrors. But at the best, the lower-SES child is likely to have peers who are on average less intellectually stimulating than those available to higher-SES children, and likely also to have to go to schools having poorer teachers, larger classes, worse facilities, and less parental involvement. It is scarcely surprising that a lower-SES environment results in lower IQs and academic achievement.

Class, Money, and the Gap between the United States and Other Developed Countries

How much would the IQ and achievement gaps be affected if the poor simply had more money? We know that the United States accepts lower levels of intellectual accomplishment for its lower-SES children than do other advanced countries. We need to consider the SES gap in achievement in light of the unusually large economic gap between classes in the United States. Income inequality in the United States is much higher than in most European Union countries or Japan. Although the income per capita in the United States is 25 to 35 percent higher than in most other advanced countries, workers in the bottom third of the income distribution are poorer than workers in the bottom third in the European Union or Japan. And workers in the bottom 10 percent of the income distribution of the average European Union country earn about 44 percent more than Americans in the bottom 10 percent. And even this statistic underestimates the disparity between the poorest Europeans and the poorest Americans. Europeans have national health insurance and other economic cushions that most Americans at low-income levels either pay for out of their own pockets or do without.

Income disparity is growing at a much faster rate in the United States than in almost any other advanced country. In 1979 the top 10 percent of wage and salary workers earned 3.5 times as much per hour as workers in the bottom 10 percent. Twenty-six years later, the top 10 percent earned 5.8 times as much as the bottom 10 percent. For families with children, the after-tax incomes of the lowest fifth rose by just 2.3 percent in the period 1979–2002. In contrast, the after-tax incomes for middle-income families with children rose by 17 percent during this same period. Between 1997 and 2006, the federally mandated minimum wage was never increased. Although an act substantially increasing the minimum wage was passed recently, the minimum wage in 2009, when the maximum is reached, will be only 73 percent of what it was in 1968 in real dollars.

Reflecting the differences in income inequality, there is more skill disparity between the social classes in the United States than there is in most advanced European countries, as measured by literacy, mathematics, and science scores gathered by the Organisation for Economic Co-operation and Development. Americans in the top fourth of SES scored almost a standard deviation higher than those in the bottom fourth. The comparable difference for the Scandinavian countries was less than two-thirds of a standard deviation. Most of the difference is due to the better performance of Scandinavians in the bottom fourth of the socioeconomic distribution. The difference in reading and math skills between lower- and higher-SES groups in the United States is greater than that for twenty-two industrialized countries that have been studied. The difference between the United States and South Korea is even more marked: only a third to a half of a standard deviation separates the average academic achievement of the bottom quarter on the socioeconomic index from that of the top quarter.

In fact, the achievement gap between the lowest 25 percent and the highest 25 percent of Americans is more similar to that in developing countries than developed countries.

There is every reason to believe that the IQ and achievement gaps in the United States could be reduced if people of lower SES had higher incomes. Low incomes produce many problems, ranging from poorer nutrition and health, to more disruption due to moving from place to place, to lowered expectations for the rewards of education. In a vicious feedback loop, lower income brings lower academic achievement to lower-status American youth, which in turn lowers their value in the labor market, which results in continued lower SES.

In a word, if we want the poor to be smarter, we need to find ways to make them richer.

Cognitive Culture

But other factors that contribute to the test score gap are not so readily cured by money. They concern the parenting practices of the lower-SES population that make learning less likely in the home and more difficult in school—cognitive culture, in short.

Higher-SES people start preparing their children for life in the fast lane early on. While their children are still in the crib, higher-SES parents begin to push them in directions that put them in good shape for the kinds of questioning, analytic minds they will need as professionals and high-level managers. Lower-SES people are not raising doctors and CEOs; they are raising children who will eventually be workers whose obedience and good behavior will stand them in good stead with employers who are not looking to be second-guessed or evaluated.

Psychologists Betty Hart and Todd Risley of the University of Kansas carried out an extremely ambitious study of the differences in verbal behavior directed toward children among white professional people, working-class blacks and whites, and underclass, welfare blacks. They observed children and their parents in their homes for many hours. In this chapter I will focus on the differences between professional and working-class families.

Professional parents talk to their children more than working-class parents do. The mother bathes her child in words, with running commentaries about the world, and about her own experiences and emotions, and with questions about the child's needs and interests. The working-class parent talks less to the child, and more of what is said is in the form of demands that would not likely stimulate the child's intellectual curiosity. The professional family includes the child in conversations at the dinner table, often attempting to engage the child in the issues that are being discussed, and exposing the child to vocabulary at the same time. Working-class parents in contrast are more likely to carry on discussions without any assumption that the child would have an interest in the topic or have anything to contribute.

The professional parent speaks about 2,000 words per hour to the child, whereas the working-class parent speaks about 1,300. By the age of three, the child in the professional family has heard about 30 million words, and the child in the working-class family has heard about 20 million. The resulting vocabulary differences are marked. By the age of three, the professional child has command of about 50 percent more words than does the working-class child.

Parents differ in how they deal with their children emotionally too, in ways that likely play a role in developing their intellectual interests and achievement. The professional parents made six encouraging comments to their children for every reprimand. The working-class parents gave only two encouraging comments per reprimand. Degree of encouragement by parents is associated with intellectual exploration and confidence on the part of the child—and the children of professional parents are way ahead of the game in this respect.

Middle-Class Parenting: Encouraging Analysis of the World

Much of what we know about social class and children's preparation for literacy and school life comes from the classic study of socialization by anthropologist Shirley Brice Heath. Heath spent many months in a town in North Carolina studying white middle-class families (all of whom had a teacher for a mother or a father), white working-class families (in most of which the father worked in the local textile mill), and black working-class families (who were mostly farm workers, mill workers, or welfare recipients). Heath literally lived with the families, observing them during all hours of the day and night and following the children to school. She found very large differences in the literacy-related activity of the three groups of children and in their preparation for elementary school. Heath's study was conducted in the late 1970s and her evidence base is just a small number of families in a particular community, but more recent studies, with a larger number and wider range of participants, found parenting practices differing across social classes in much the same way that Heath reported. In what follows, I rely primarily on Heath's work and the more recent work of Annette Lareau.

The middle-class parent reads to the child much more than does the working-class parent. There are lots of children's books in the middle-class home. Reading to the child begins as early as six months, as soon as the child can be propped up to look at a book. And the middle-class parent reads to the child not just as a form of entertainment but also to encourage connections between what appears on the page and what exists in the outside world. There is a deliberate effort to take what is read in books and relate it to objects and events in daily life and in the world. ("Billy has a black doggie. Who do you know who has a black doggie?" "That's a robin. Where did we read about robins? What do robins eat?"). Parents also encourage analysis of what is read ("What will happen next? What does she want to do? Why does she want to do that?").

From a very early age the middle-class child expects to be asked questions about books and knows how to answer them. Parents ask their children about the attributes of objects and teach them how to categorize objects based on their properties. (I once sat on a plane behind a father and his three-year-old son. The father had a picture book and was asking the child whether particular objects were long or short. "No, Jason, pajamas are *long*.") Middle-class parents also ask *what* questions ("What's that?" "What did Bobby try to do?") and follow them with *why* questions ("Why did Bobby do that?"), and later with requests for evaluations ("Which soldier do you like better?" "Why do you like him better?"). Adults encourage their children to talk about what is in their books and even to tell stories that are inspired by the ones they have read.

Middle-class children are well prepared for school. They know how to take information from books, they expect to be entertained by them, and they are familiar with how to answer so-called known-answer questions—that is, questions whose answer is known to the questioner. The early grades go easily for such children. They are also more than ready for the later elementary years, when analysis and evaluation are called for.

Working-Class Parenting: Socialization for the Factory

The working-class baby is brought home to a house with some children's reading material—Little Golden Books and perhaps some Bible stories, maybe a dozen books all told. Walls are decorated with pictures depicting nursery rhymes, and there is probably a mobile. Family, friends, and neighbors talk to the child.

Although working-class children are asked questions about what is read to them, there is not much effort to connect what is on the page with the outside world. A book might have a picture of a duckling, and the mother might ask the child if he remembers the duck he saw at the lake, but then she might not explain the connection between the fuzzy yellow duckling on the page and the full-grown mallards at the lake. After about the age of three, children are not encouraged to carry on a dialogue with the reader. Instead they hear, "Now you've got to learn to listen." The child is supposed to pay attention, and comments or questions are regarded as interruptions.

(A Philadelphia study illustrates both a symptom and a cause of the social-class difference in literacy. In areas where almost all adults are college-educated, booksellers had 1,300 children's books available per 100 children, whereas in blue-collar Irish and Eastern European neighborhoods only 30 children's books were available per 100 children. There could scarcely be a more stark set of figures capturing the social-class literacy gap.)

Activities in the middle-class family are guided by words. The middle-class father showing his child how to bat a baseball says, "Put your fingers on top of each other around the bottom of the bat; keep your thumb in this position here; don't hold it above this line; don't leave the bat on your shoulder—hold it above your shoulder a couple of inches." The working-class child gets no such elaborate instructions or experience in how to translate verbal instructions to physical practice. Instead the child is simply told, "Do it like this; no, like this." The middle-class family, when starting to play a new game, reads the instructions aloud and comments on them. The working-class family is more likely to guess at how to play the game and start playing it, making up rules as they go along. The middle-class mother works from a recipe, which she may read out loud so her child can make connections between what is read and what materials are being used and which procedures are being carried out. The working-class mother is less likely to use a recipe, and unlikely to give her child an opportunity to make connections between it and the materials at hand when she does use one.

Working-class children come to school with sufficient preparation to do reasonably well in the early years. They often know the alphabet; they can name colors and numbers and they can count; they can tell someone their address and their parents' names. They can sit still and listen to a story, and they know how to answer *what* questions about factual matters. But when they are asked, "What did you like about the story?" not many have ready answers. When asked, "What would you have done?" they are usually stumped. When categorization and analysis and evaluation are emphasized in the later elementary grades, such children are at a decided disadvantage. When they are asked to write a story, they are likely to merely repeat some story they have been read. When asked about counterfactuals—"What would have happened to Billy if he hadn't told the policemen what happened?"—they are at a loss.

Children who face these difficulties are likely to be demoralized and alienated by junior high school and are on their way toward being candidates for dropping out of high school.

The differences between the social classes that Heath found in socialization for literacy and school helps us to understand what happens to children's IQs and academic skills over the summer, when they are not in school. The IQs and skills of middle-income children generally stagnate during this time. But there is a drop in skills for lower-SES children, whose families would not be expected to provide the degree of cultural stimulation over the summer that middle-class families do. The middle-class kids do not fall behind much during the summer because, undoubtedly, they engage in more educationally valuable activities, like reading and being read to, listening to stimulating conversation at the dinner table, going to museums and zoos, and taking classes in art, music, and even academic subjects. One study found that of children who are in transition between kindergarten and first grade, those in the highest quintile of SES actually show an increase in skills over the summer, whereas kids in the lowest SES quintile show a substantial decrease. So a significant portion of the difference in IQ and academic skills between upper-middle-class children and lower-class children can be attributed to the cumulative drop over the summer for lower-SES kids, which they never quite make up for during the school year.

The hereditarian reading the chapter to this point might be thinking, "How do we know that these differences in socialization practices actually play a causal role in the intelligence and achievement of children? How do we know that it isn't simply that higher-SES people have more intelligent children than lower-SES people, not just because of what the environment does to them, but simply because they have their parents' fortunate genes? And parents with smarter genes do intellectually stimulating things for their kids all right, but this is because their genes make them enjoy doing those things and the children are more rewarding to do those things with because they're smarter."

Undoubtedly, what the hereditarian says accounts for a nontrivial portion of what is going on. The environmental differences are to some degree a consequence of higher-IQ genotypes from upper-SES parents and lower-IQ genotypes from lower-SES parents.

But remember that genetics cannot be a very large part of the explanation of the IQ and achievement gap. The purely environmental contribution to the gap between the highest- and lowest-SES groups (probably roughly the highest and lowest 15 percent) is 12 to 18 points. That does not leave a lot of room for genes. The gap between the lower third in SES and the upper third is 10 points, and we know that a substantial portion of that has to be due to environmental differences.

Our confidence about the very substantial role of the environment is crucial to keep in mind for when I discuss in a later chapter how much we might expect to improve the intelligence of working-class and lower-class children. Both because of the adoption studies discussed earlier and because of the litany of environmental factors that I have recited in this chapter, we know that there is lots of room for improvement in the environment to make a difference.

Improving the economic situation of the poor would undoubt-edly make a big difference. On the other hand, there would not necessarily be big improvements in the IQ and academic achieve-ment of lower-SES children in the first generation after economic advancement, if that were to take place. Generals prepare for the last war, and parents socialize for the life situations of their parents and not so much for their own or their children's. In fact, there is some evidence that in the first generation after additional income is made available to families, the improvement in intellec-tual status of the children is slight. Gains due to improvement in income alone are likely to be incremental across generations.

Fortunately, as you will soon see, the schools can be made to speed up the process of reducing the gap between the social classes, putting lower-SES children in a much better position to take advantage of the opportunities available in the information-age economy.

But first, let's look at the gaps between the races in IQ and aca-demic performance. Some of the reasons for the gaps are the same as those for the SES gaps, and some are different.

CHAPTER SIX

IQ in Black and White

The taboo against discussing race and IQ has not left this an open question. On the contrary, it has had the perverse effect of freezing an existing majority of testing experts in favor of a belief that racial IQ differences are influenced by genetics. No belief can be refuted if it cannot be discussed.

—Thomas Sowell (1994)

[Black] kids seem to . . . have . . . this unconscious way of thinking that Blacks are inferior to Whites. And I think that takes a toll. —Black male high school student in Ohio interviewed by anthropologist John Ogbu (2003)

THE QUESTION OF whether there are innate differences in intelligence between blacks and whites goes back more than a thousand years, to the time when the Moors invaded Europe. The Moors speculated that Europeans might be congenitally incapable of abstract thought!* But by the nineteenth century most Europeans probably believed that they were genetically superior to Africans in intellectual skills.

The IQ test, developed early in the twentieth century, reinforced the genetic view. Since whites scored higher than blacks, many psychologists, basing their hypothesis on the assumption that IQ is largely heritable, assumed that the black/white group differences were genetic in origin.

For decades, whites scored an average of 100 points on IQ tests, while blacks scored about 85—a difference of 15 points or

---

*A millennium earlier southern Europeans had their doubts about northern Europeans. Cicero warned the Romans not to purchase the British as slaves because they were so difficult to train, though Julius Caesar felt they "had a certain value for rough work."

a full standard deviation. If such a difference were wholly or substantially genetic in origin, the implications for American society
would be dire. Even if the environmental playing field were lev-
eled, a much higher proportion of blacks than whites would have
trouble supporting themselves, and a much lower proportion of
blacks than whites would be capable of being successful in business or the professions.

In this chapter I review the evidence about the role of genetics in producing differences in IQ between blacks and whites. I
also show that there remain serious societal roadblocks, as well
as some social practices characteristic of African Americans, that
make educational and occupational advancement less likely.

Not in the Genes

Laypeople differ markedly in whether they think race differences
in IQ have a partly genetic origin or a purely environmental
origin—and so do behavioral scientists. Some laypeople I know—
and some scientists as well—believe that it is a priori impossible
for a genetic difference in intelligence to exist between the races.
But such a conviction is entirely unfounded. There are a hundred
ways that a genetic difference in intelligence could have arisen—
either in favor of whites or in favor of blacks. The question is an
empirical one, not answerable by a priori convictions about the
essential equality of groups. As it turns out, there is a great deal
of empirical evidence on the question.

In 1994, psychologist Richard Herrnstein and political scientist
Charles Murray published *The Bell Curve,* which maintained that
black IQs were clearly lower than white IQs. They presented what
they maintained was a balanced view of evidence pointing to the
conclusion that the IQ difference between the races was substantially genetic. Charles Murray—Herrnstein died at the same time
that *The Bell Curve* was published—has repeatedly claimed that
the book was agnostic with respect to the degree of genetic contribution to the difference. The evidence presented in the book,

however, is clearly weighted in favor of the genetic view. Both the public and most of the scientific community have concluded that the book endorsed the view that the black/white difference was probably substantially genetic in origin.

In this chapter and in Appendix B I present the arguments for genetic determination of the black/white gap made by Herrnstein and Murray in their book, and by J. Philippe Rushton and Arthur Jensen in their sixty-page review of the evidence published in 2005. The appendix is not intended for the general reader but rather for the specialist who would like to see the entire case against the genetic argument.

The black/white gap is not due to some obvious artifact, such as blacks not being familiar with formal English, or being less motivated to perform on IQ tests, or having teachers or IQ testers who have low expectations for their performance.

There is plenty of evidence though, that blacks sometimes perform worse on IQ tests and achievement tests when their race is made salient and this engages a "stereotype threat," causing them to perform worse than they would in more relaxed settings where they are not afraid of confirming a stereotype that white testers have. Social psychologists Claude Steele and Joshua Aronson demonstrated this, and countless subsequent studies have confirmed the point. The underperformance is most likely when blacks are tested in an integrated setting and it is made explicit that it is intellectual ability that is being tested. When the test is presented as a puzzle, or when blacks are assured that blacks and whites do equally well on the test, black performance typically is better than in more threatening circumstances, sometimes markedly better.

In general, it is not the case that blacks perform better either at school or at work than would be indicated by their IQ scores. At least as late as 1980, when Jensen reviewed the question, academic performance and occupational outcomes for blacks were actually lower than would be predicted by their IQ scores. At a given IQ level, whites perform better than blacks.

Blacks have lower socioeconomic status (SES) on average, and people with low SES have lower IQ test scores. But that fact by itself does not speak to the heritability issue, because it is not clear to what extent low SES drives IQ lower versus to what extent low IQ drives SES lower. We do know that blacks have lower IQs than whites at every level of SES, so SES cannot be the full explanation of the black/white IQ difference.

The hereditarians have come up with a raft of evidence that the black/white difference in IQ is genetically based. But all such evidence is indirect. There is, for example, the evidence that brain size is associated with intelligence. The correlation between brain size and IQ may be as high as .40. And according to a number of studies by Rushton, blacks have smaller brains than whites.

The correlation between brain size and IQ does not indicate a causal relationship, however. If bigger brains were smarter because of their size, then we would expect to find a correlation within families. Siblings who get the larger brains by luck of the genetic draw should also be the ones who have higher IQ scores. In fact, however, there is no such correlation.

Moreover, the brain-size difference between men and women is substantially greater than that between blacks and whites as reported by Rushton and Jensen, yet men and women score the same, on average, on IQ tests. And a group of people in a community in Ecuador have a genetic anomaly that produces extremely small head sizes—and hence brain sizes—yet their intelligence is as high as that of their unaffected relatives, and their academic achievement is substantially greater than that of most people in their communities. The direction of recent evolution over the last few thousand years, incidentally, is toward smaller brain sizes for humans. And I note just for interest's sake that Albert Einstein's brain was decidedly smaller, at 1,230 grams, than the overall average found for blacks in the studies by Rushton.

Most of the evidence that the hereditarians present is indirect in the same way that the brain-size evidence is. It is not necessary, however, to rely on indirect findings when we have much more direct evidence about the basis for the IQ gap. A natural experiment allows us to test whether European genes make for higher IQs than African genes. About 20 percent of the genes in the American black population are European, meaning that the genes of any individual can range from being 100 percent African to mostly European. If European genes for intelligence are superior, then blacks who have relatively more European genes ought to have higher IQs than those who have more African genes.

One test of this hypothesis is to determine whether the physical features of blacks that indicate more European heritage are associated with higher IQs. It turns out that light skin color and stereotypically Caucasian features—both measures of the degree of a black person's European ancestry—are only very weakly associated with IQ (correlations in the .10 to .15 range), even though we might well expect a moderately high association due to the social advantages of these characteristics.

Another test of the genetic hypothesis occurred as a result of World War II, when both black and white American soldiers fathered children with German women. Thus some of these children had 100 percent European heritage, and some had substantial African heritage. Tested in later childhood, the German children of the white fathers had an average IQ of 97.0, and those of the black fathers had an average of 96.5, a trivial difference.

If European genes conferred an advantage, we would expect the smartest blacks to have substantial European heritage. But when a group of investigators sought out the very brightest black children in the Chicago school system and asked them about the race of their parents and grandparents, they found that these children had no greater degree of European ancestry than blacks in the population at large.

Blood-typing tests have been used to assess the degree to which black individuals have European genes. The blood group assays show no association between degree of European heritage and IQ. Similarly, the blood groups most closely associated with high intellectual performance among blacks are no more European in origin than other blood groups.

One way of testing the heredity-versus-environment question is to look at black children raised in white environments. If the black deficit in IQ is due entirely to the environment, then blacks raised in white environments ought to have higher IQs than those raised in black environments. The hereditarians cite a study from the 1980s showing that black children who had been adopted by white parents had lower IQs than white children adopted by white parents. Mixed-race adoptees had IQs in between those of the black and white children. But, as the researchers acknowledged, the study had many flaws; for instance, the black children had been adopted at a substantially later age than the mixed-race children, and later age at adoption is associated with lower IQ.

A superior adoption study was carried out by developmental psychologist Elsie Moore, who looked at black and mixed-race children adopted by middle-class families, either black or white, and found no difference in IQ between the black and mixed-race children.

Important recent research helps to pinpoint just what factors shape racial differences in IQ scores. Psychologists Joseph Fagan and Cynthia Holland tested black and white community-college students on their knowledge of, and their ability to learn and reason with, words and concepts. The whites had substantially more knowledge of the various words and concepts, but when participants were tested on their ability to learn new words, either from dictionary definitions or by learning their meaning in context, the blacks did just as well as the whites.

Whites showed better comprehension of sayings, better ability to recognize similarities, and better facility with analogies when solutions required knowledge of words and concepts that were more likely to be known to whites than to blacks. But when these kinds of reasoning were tested with words and concepts known equally well to blacks and whites, there were no differences.

Within each race, prior knowledge predicted learning and reasoning, but *between* the races it was only prior knowledge that differed, not learning or reasoning ability.

It seems unlikely that differences in knowledge would have a genetic basis if there are no differences between the races in learning and reasoning ability. It seems much more likely that the knowledge differences are entirely due to environmental effects. (However, I would never argue that knowledge differences do not count as intelligence differences. Intelligence depends to a substantial degree on knowing words and concepts.)

The Fagan and Holland research is extremely important, but replication with different kinds of materials and different participants is in order before we can be completely confident about its implications.

Some of the most convincing evidence about whether the IQ gap has environmental causes concerns Flynn's discovery about IQ changes over recent generations. This research, described in Chapter 3, established that in the developed world as a whole, IQ increased markedly from 1947 to 2002. In the United States alone, it went up by 18 points. Genes could not have changed enough over such a brief period to account for the shift; it must have been the result of powerful social factors. And if such factors could produce changes over time for the population as a whole, they could also produce big differences between subpopulations at any given time. Indeed, black IQ now is superior to white IQ in 1950. If black genes for IQ are inferior to white genes for IQ, that could not happen—unless you wanted to argue that the environment for blacks today is far more conducive to high IQ than the environment for whites in 1950. I doubt that many people would attempt to make that argument.

Finally, since there is good reason to believe that the environment of blacks has been improving at a more rapid rate than that of whites, the black/white gap should be less today than in the past. In fact, we know that the IQ difference between black and white twelve-year-olds has dropped to 9.5 points from 15 points

in the last thirty years—a period that was more favorable for
blacks in many ways than the preceding era. Black progress on the
National Assessment of Educational Progress (NAEP) Long-Term
Trend test shows equivalent gains. Reading and math improvement has been modest for whites but substantial for blacks. The
shrinkage of the gap on the NAEP is roughly equal to the 5.5
points found by Dickens and Flynn for IQ tests.

It is hard to overestimate the importance of a gap reduction
of this magnitude. When population means differ, the differences at the tails of the bell curve are very great. If whites score
higher than blacks by an average of 15 IQ points—the difference between the two groups in the past—people with IQs of
130 and above are about 18 times more likely to be white than
black. That would mean that successful doctors, scientists, and
professionals would be enormously more common among whites
than among blacks. But if the difference is 10 points, the ratio of
whites to blacks is more on the order of 6 to 1. Whites would be
much more common at the highest levels of achievement, but the
difference between the two groups would not be as great. Differences at the other end of the IQ scale are equally marked. If the
group difference is 15 points (blacks scoring lower than whites),
people who will not be able to fend for themselves economically
are vastly more likely to be black than white; if the difference is
10 points, such people are merely substantially more likely to be
black than white.

Barriers to Accomplishment for African Americans

So why is it that blacks historically score poorly on IQ tests,
achieve low levels academically, and attain relatively low levels of
occupational success? The evidence indicates that genes play no
role in these facts. So we can assume we have a purely environmental story to tell.

First, all of the problems of lower-SES people that affect ability
and achievement are often exacerbated for blacks, who are overrep-

resented among the poor. To remind you of just what those potential problems are: They include poor prenatal care and nutrition,
relative infrequency of breast-feeding, hunger, deficiency of vitamins and minerals, lead poisoning, fetal alcohol poisoning, poorer
health care, greater exposure to asthma-causing pollution, emotional trauma, poor schools, poor neighborhoods along with the
less desirable peers who come with the territory, and much moving
and consequent disruption of education. For the black underclass
these problems are worse than they are for poor whites.

Additional problems also exacerbate the black situation. Black
family income in 2002 was 67 percent of white family income, but
black family wealth was 12 percent of white family wealth! Part
of the reason for this discrepancy is the practice of "redlining"—
keeping blacks out of white neighborhoods where property values
produce a greater return on investment. Thus for blacks in general, there is very little cushion to fall back on in times of under-
employment or unemployment. For lower-class blacks there is
next to nothing. Moreover, the wealth differences reflect the fact
that most blacks who are in the middle class have only just arrived
there. We can expect that their parenting practices are going to
remain more similar to that of the lower class than is the case for
the typical middle-class white.

The unwed mother rate is 72 percent for blacks, compared to
24 percent for whites. This statistic represents a host of problems
for black children, not least of which is that the poverty rate for
single-parent homes is far higher than it is for two-parent homes.
Perhaps equally important is the fact that such homes have only
one adult, and the fewer adults there are, the less stimulating is
the environment.

The economic situation for blacks is not static. Two different
trends are developing in the black community in the United States,
one favorable and one highly unfavorable.

First, more and more blacks are moving into the middle class,
and those already there are strengthening their economic circumstances. Affirmative action has likely played a role in these gains

(though some researchers argue that substantial progress occurred for blacks before affirmative action became common).

A second trend is that the economic situation is getting relatively worse for blacks who remain in the lower class, just as it is for lower-SES people in general. Recall from the last chapter that real income is lower for the poor and for working-class people than in the past. For lower-SES blacks, affirmative action has had little effect. So the financial situation is worse for a large portion of the black population, especially young black males, than it was in the past.

The motivation to work, particularly for young black men, is undercut by societal attitudes that this group is not to be trusted. Employers believe that young black men are less dependable, enthusiastic, cooperative, and friendly, are less capable of working with teams, and have poorer communication skills.

Unfortunately, there is evidence that employers are not capable of setting this stereotype aside even when presented with an individual who clearly contradicts it. When black and white job applicants with similar credentials apply for jobs, the white applicants do much better. An experiment with black and white college graduates posing as applicants for entry-level jobs is particularly chilling. The young men, all of whom were well groomed and articulate, presented themselves as high school graduates with identical qualifications. The white applicants *who admitted to a felony conviction* got more positive responses than did blacks with no such blot on their records.

So powerful stereotypes brand even the black male who is a high school graduate. This means that, realistically, a diploma has less value for black males than for white males. It would scarcely be surprising if this lowers the incentive for black males to complete their education. It makes sense, then, that black females do far better educationally than black males, and consequently reach higher occupational levels. Already in 1965, at the time of Patrick Moynihan's well-known report on the status of black families, black females were 30 percent more likely to graduate from high school than black males. In 2005, for blacks twenty-five to twenty-nine years old, the ratio of females to males with a college degree was 1.69 to 1.

Since we know that more schooling and more serious attitudes toward school make people smarter, it comes as no shock that as of about 1980 black females were twice as likely to have an IQ above 120 than were black males. There is of course no conceivable genetic explanation for such gender differences, given that there is no mean difference in IQ between white males and females and given that, for whites, males are more represented at the high end of the IQ distribution (as well as at the low end, which is how the average can come out to be the same). As a consequence at least in part of their superior educational credentials, black women are twice as likely to have white-collar jobs in the federal government as are black men.

So not only must lower-SES blacks face all the hardships of their social class, but also they face a separate host of problems owing to their race. SES is worse for blacks than for whites, and prejudice is devastating for large swaths of the black population, reducing their ability to get decent jobs and sapping their motivation to complete their education.

The Caste System in America for Blacks and "Former Blacks"

Race may be an "American dilemma," as the title of a classic work by Gunnar Myrdahl suggests, but the problems of lower castes throughout the world are similar to those of American blacks. African anthropologist John Ogbu has reviewed evidence showing that caste-like minorities such as the Maori in New Zealand, the Burakumin in Japan, Catholics in Northern Ireland, Sephardic Jews in Israel, and scheduled castes ("untouchables") in India are characterized by poor school performance, high likelihood of dropping out of school, low IQ test scores, and high levels of crime and delinquency. The IQ differences between higher and lower castes in India exceed the largest differences reported between blacks and whites in the United States. Moreover, many non-African groups in the world have recorded average IQs of 85 or lower—less than the value commonly given for blacks in the United States in the past and substantially less than the contemporary value for blacks. These groups include whites in some Appalachian Mountain communities in the early-twentieth century, the children of the first wave of Italian immigrants to the United States, canal boat communities in Britain, and inhabitants of the Hebrides islands off Scotland.

Ogbu focuses on what he calls "involuntary" minorities, those like American blacks who were brought by force to America. He contrasts them with what he calls "autonomous" minorities, whose members are separate from society by choice, such as Mormons and the Amish, and with immigrant minorities. Immigrants tend to compare themselves not with the majority in the new country but with their peers in the old country, and they find themselves better off. Unlike lower-caste minorities, immigrant minorities tend to have lower rates of crime than the host population. (Though their children have higher crime rates, probably in part because they *do* compare their situation unfavorably with that of the majority in the society.)

Ogbu holds that caste-like minorities often fail to take full advantage of the opportunities that are available to them because they do not have "effort optimism"—that is, they lack conviction that their efforts will be rewarded. A frequent consequence of this lack of faith in the system is that they do not work hard in school because they do not expect the work to be rewarded. Younger members of the minority may even invert the educational values of the society. In the case of American blacks, the young may reject academic earnestness as "acting white." Ogbu has written for decades about the attitudes of young blacks, especially males. Young black students, even middle-class ones in middle-class schools, are more likely to come to class without having done their homework, and they are more disruptive. A surprising number think they can go to college on athletic scholarships and so they do not need to make good grades. They are likely to avoid taking academically challenging courses even when encouraged to do so by teachers and counselors.

Knowledge of the cultural history of blacks in the United States is important to understanding race problems today. Here I review some of that history and compare it with the history of another group that was formerly the target of great prejudice— Irish Americans. In the nineteenth century in the North it was far from clear whether the Irish or the blacks would be better off by the twenty-first century. In what follows I draw especially on the work of Thomas Sowell and James Flynn.

There was a black presence in America from the very beginning of European settlement. Twenty black indentured servants landed in Jamestown twelve years after the founding of the colony. In eighteenth-century Virginia, blacks and whites were as integrated as they are today, and in some ways more so. Blacks and whites commonly attended the same church, and it was not unusual for the minister to be black. The European settlers adapted many aspects of African culture, including agricultural methods, cuisine, and mythology, to the common civilization that the two groups were building together.

Blacks in both the North and the South in the eighteenth century had, if not an equal place at the table, at least an accepted and valued status in the larger society. The lives of most blacks were not all that different from those of white indentured servants, who served for a term as apprentices and then were free to do as they chose. Blacks were not bound for a lifetime to a particular master.

From early on, many blacks in the North had never been slaves or had been freed from slavery. In the big cities of the North, "free persons of color" were preferred to the Irish as employees and neighbors right up to the early-twentieth century, when it was still possible to see signs at factories and in shop windows reading, "Colored man preferred. No Irish need apply." Though most free blacks in the North were servants and semi-skilled or unskilled workers, their numbers included artisans, tradesmen, and skilled workers in the early years. In Boston in 1860, blacks had a higher occupational status than the Irish, and hotels in New York paid black employees higher wages than they paid the Irish.

Eighty-five percent of free black families dwelling in both the North and the South in the 1855–80 period were headed by males. In Philadelphia, male-headed families were more common among free blacks than among any other group. The percentage of families that were headed by males was much lower for the Irish in the nineteenth century than it was for blacks. Racial segregation was not substantial in the cities of the North until well into the twentieth century. In Chicago in 1910 more than three-quarters of blacks lived in neighborhoods that were mostly white.

To give an idea of how different many of the urban free persons of color were from the image of the urban black of today, consider the group of five hundred free blacks in the Washington, D.C., of 1800 and their descendants. They created their own schools, beginning in 1807, which black children attended until they were finally admitted to public schools in 1862. They also founded the first black high school in 1870. From that time until the middle of the twentieth century, three-quarters of the students in that high school went on to college, above the average for whites even today. In the early 1900s, students at the black high school scored higher on a city-wide achievement test than did students at any white high school in D.C. Once IQ tests began being given, students in that high school scored above the national average. Its graduates include the first black general, the first black cabinet member, the first black federal judge, the first black senator since Reconstruction, and the discoverer of blood plasma.

It was not foreordained that the free persons of color would fail to become fully equal citizens in the North. They had a head start on the Irish. It was the enslavement of large numbers of blacks in the South and the subsequent migration of large numbers of impoverished and illiterate blacks to the cities of the North beginning in the late-nineteenth century that undid the northern blacks.

Slavery created conditions for blacks in the South not unlike those for the Irish in Ireland. A crucial fact for both groups is that labor did not produce value for the worker, and so hard work was not initially a cultural value for either group. The work of the slave resulted in gain only for the slave owner, and the work of the Irish resulted in gain only for the absentee English landlord. Even the cottage—or more typically hovel—where the Irishman lived was owned by the landlord, so the Irishman had little inducement to make improvements to the property because they would only redound to the advantage of the owner. An ecological fact about Ireland led in a direct way to still another reason for the traditional Irish dislike of work: the most productive use of the soil in Ireland was for the growing of potatoes, but the crop requires only a couple of weeks of work a year. When the Irish first came to the New World in large numbers in the mid-nineteenth century, they had no tradition of steady work. More than a century would pass before their reputation as layabouts was put to rest.

The end of slavery in the South did not result in economic freedom for the blacks. The conditions of the blacks under slavery were superior in many ways to what they were to become in the post–Civil War period. Although blacks initially enjoyed a measure of political freedom during the Reconstruction period—and indeed many were elected to federal office because former rebels were not allowed to vote or run for office—huge numbers of blacks subsequently were forced into peonage in the form of sharecropping. By the latter part of the nineteenth century, after the imposition of Jim Crow practices of discrimination, economic conditions were poor for the great majority and desperate for many. In an attempt to escape poverty and discrimination in the South, many blacks fled to northern urban centers. The great migration to the North began at the end of the nineteenth century and roughly doubled each decade until 1940.

The people who arrived in the North were desperately poor, had few skills and little education, and were thoroughly rural in habits and attitudes. They overwhelmed the descendants of the free persons of color, changing the nature of the indigenous black community and bringing considerable social pathology to the cities of the North.

If the labor market had been fully open to the new arrivals, the history of blacks in the urban North in the twentieth century might have been a thoroughly positive one. But the labor movement for the most part excluded blacks. Black men were denied entry into unions just as access to high-paying jobs became ever more dependent on holding a union card. Only in a few industries in a few regions, such as the auto industry in Detroit, were blacks allowed into the unions in significant numbers. Elsewhere blacks were relegated to semi-skilled and unskilled jobs, and even for those jobs blacks were the last hired and the first fired.

The Irish, who were white (or as author Noel Ignatiev put it, became white), were gradually admitted into unions and rose from the underclass to the lower class and ultimately in great numbers into the middle class. The two other institutions that had a great deal to do with the rise of the Irish were politics—and the patronage that came with it—and the Catholic church, which waged a heroic struggle to educate poor Catholic immigrants.

As of the mid-twentieth century, the Irish in Ireland had IQs at about the level of blacks in America. English psychologist H. J. Eysenck attributed this to the genetic consequence of the fact that the intelligent people had fled Ireland to other lands, leaving the dull-witted—and their inferior genes—behind. The gene pool of Ireland must have been more robust than Eysenck thought, however, because the per capita gross domestic product of Ireland is now greater than that of England, and literacy proficiency for children is higher than that of the United Kingdom. (This achievement was no accident. It was the result in part of an intensive education initiative begun in the 1960s. Post–secondary school enrollment increased from 11 percent in 1965 to 57 percent in 2003.)

Despite the dislocating effects of the black migration from the South on both the immigrants and the cities they came to, and despite the near-exclusion of the immigrants from high-paying union labor, economic conditions for the immigrants were far better than they had been in the South. The conditions of the urban poor improved steadily throughout the century, with the exception of the Depression era. A huge spurt in economic conditions came in the 1960s and 1970s (perhaps not coincidentally, the period during which the children who made great gains in school achievement were born). By 1970, black families with a husband and wife who both worked made almost as much money as comparable white families.

The size of the black middle class continues to expand. The proportion of blacks who are occupationally in the middle class moved from 10 percent in 1950 to 31 percent in 1976 to 52 percent in 2002. But already by the 1960s, the fortunes of the black population were bifurcating. The middle class was growing, but large numbers of blacks were remaining mired in deep poverty.

The economic divide in the black community has everything to do with the difference that stable marriage makes. The households where a man is present and employed are doing very well. But the roughly two-thirds of black families headed by a woman are doing much less well. For every black man who drops out of school and hence often out of the workforce, there is likely to be a single mother who must fend for herself. Every female-headed family is more likely to produce males who will drop out of school and out of the workforce and out of the marriage pool. And so on, in a vicious cycle.

Caribbean Cultural Capital

In addition to the descendants of the northern free blacks and the former slaves, there is a third distinctive group of blacks. It has a very different, and very grisly, past but a much better present and a brighter future. This group is the West Indian immigrants to this country. West Indians constitute less than 1 percent of the population but have produced a vastly disproportionate number of eminent black Americans—from Marcus Garvey to Colin Powell. In 1970, second-generation West Indians exceeded Americans in general in median family income, in level of education, and in percentage in the professions. Unlike black immigrants from the South, West Indian immigrants to the big cities of the North behaved like many other groups of foreign immigrants. They took whatever jobs were available, saved their money, started or purchased small businesses, and saw to it that their children were educated even if they had to forgo luxuries or necessities for themselves. They moved into the professions and high-ranking business positions at a far higher rate than did native blacks.

On the surface, these accomplishments are particularly remarkable given that the nature of slavery in the West Indies was in every respect more gruesome and inhumane than that of the United States. Nor can the success of West Indians in America be traced to a greater dose of either European genes or European culture. The admixture of European genes to the West Indian pool was far less than it was in the United States. And West Indian culture is much more rooted in African culture than is American black culture. (Sowell recently argued that the culture of the "street" people in the black inner city owes much more to the culture of Northern Ireland and the Scottish borderlands of the eighteenth century, and to the white rednecks who modeled that culture for the blacks in the U.S. South, than to the West Africa of that period.)

In Sowell's view, the key to West Indian exceptionalism lies in the economic history of slavery in that part of the New World. In the U.S. South, slaves were fed from a central kitchen or had food doled out to them for preparation in their own quarters. In the West Indies, blacks raised their own crops and sold the surplus in the markets. Because of the small number of whites, all of the skilled labor and artisanship as well as much of the entrepreneurship had to come from the black population. The West Indians arrived in this country poor, but they were in a far better position to take advantage of the opportunities in America than were their country cousins from the U.S. South.

I must point out that West Indian immigrants are not a random sample of the West Indian population as a whole. The very lowest echelons of West Indian society have not been well represented among the immigrants, who include a much higher proportion of the professional and managerial classes than is characteristic of the West Indian population as a whole. Self-selection, in short, counts for some unknown portion of the success rate of West Indians in this country, and this self-selection undoubtedly includes some unknown degree of genotypic advantage for intelligence.

Despite this disproportionate influx of skilled individuals, the occupational and educational success of West Indians tells us something crucial about the role played by racism in the lower occupational achievement of blacks. However severe racism may be, it does not prevent blacks from attaining high levels of achievement if they have good skills and favorable attitudes toward work. (Of course, it helps that in New York City and other places where there are large numbers of West Indian blacks, the stereotype of West Indians is that they are reliable and hard working. A lilting Caribbean accent is an asset on the labor market.)

Parenting Practices

The cultural capital of West Indians is unusually great, whereas that of some native blacks is distinctly lower than that of the majority population. On top of all the demographic disadvantages of American blacks as a whole, many also socialize their children in ways that are less likely to encourage high IQ scores and high academic achievement than do whites of comparable social and economic circumstances.

Nothing is more telling than the way that many black parents interact verbally with their children. I pointed out in the previous chapter (on social class) that children of professionals hear about 2,000 words per day and working-class children, both black and white, hear about 1,300. Children born to blacks on welfare, however, hear only about 600 words per day. By the time the child of professionals is three years old, she has heard 30 million words; the child of welfare blacks has heard 10 million. The vocabulary used by the three-year-old child of professionals in talking to her parents is richer than that of the black welfare mother in talking to her child.

Recall from the last chapter the study by anthropologist Shirley Brice Heath of children in a rural community in the South. During the 1960s and 1970s, she lived with three different types of families for a period of many months and followed their children to school. She studied white middle-class families in which at least one parent was a teacher, white working-class families, and black lower-class families. Many of the respects in which working-class families differed from middle-class families were found to be characteristic of poor black families as well. But many aspects of black socialization for language were quite different from the practice of either class of whites—and even more disadvantageous for preparation for school.

The poor black child was born into an extended family that bathed it in communication, both verbal and nonverbal. But the adults did not speak to the child directly—they made no effort to interpret the baby's sounds as words. They did not simplify their language for the child. Nor did they label objects or events or make any attempt to link objects in the here and now with other objects encountered in other contexts. In other words, they did not decontextualize things in such a way that learning in one situation could be carried over into others.

Children didn't play with educational toys but with safe household objects instead—spoons, plastic food containers, pot lids. Older children might get electronic and mechanical toys. But they didn't get manipulative toys—blocks, take-apart toys, or puzzles—to play with.

Nor did the children get books. The adults read newspapers, mail, calendars, advertising brochures, and the Bible, but there were no special reading materials for children other than, sometimes, Sunday-school materials. The adults did not sit their children down and read to them. There was no special bedtime ritual, or even a specified time for going to sleep when the bedtime story might become a fixture.

The children were not asked *what* questions about their environment. They were instead asked for nonspecific comparisons: "What's that like?" (This probably pays off later in an ability to see similarities. Similarities subtests of IQ tests are the ones that blacks do best on.) The children's abilities to link two events together metaphorically produced no advantage for them in school. In fact, those abilities often caused problems because they allowed children to see linkages that the teacher had not intended. By the later grades of elementary school when children are asked to compare and evaluate—and when the skills in detecting similarities would have been useful—the children had fallen too far behind. They did not have the verbal and written comprehension skills that would allow them to put analogies in a form that teachers might accept.

In the home, the children were not asked known-answer questions—that is, questions for which the adult knows the answer ("What color is the elephant, Billy?"). As a consequence the children were not prepared for such questions when they started school. Even the simplest question from the teacher might go unanswered because the child was nonplussed by the form of the question ("If the teacher doesn't know this, then I sure don't.")

The children did engage in storytelling in the home—if they could gain center stage for long enough. But the stories told were not likely to be impressive in the school context. They typically had no beginning and no ending—just an effort to entertain until the audience lost interest. The children had narrative abilities that exceeded those of white working-class kids and even those of many if not most white middle-class kids. (The narrative skills of blacks are much in evidence in the entertainment industry and among clergymen. There is a saying that the worst black preacher on his worst Sunday is a more effective speaker than the best white preacher on his best Sunday.)

In school, children were asked to classify objects with respect to shape, color, or size. But the impoverished black children had not been taught to categorize, so this exercise was very foreign to them. And when they were asked to interpret scenes in books—which are inevitably stylized—they often found it difficult to relate the scenes to concrete objects and events in the world.

In the late 1980s, Heath returned to North Carolina to study the children of the children she had studied more than twenty years before. The original community she had visited no longer existed. The clothing mills where some of the parents had worked had closed, and the farms were now mechanized. While some of the children had escaped to middle-class occupations in the cities, the people Heath studied were those who lived in a slum area in the original town and others who lived in a low-income high-rise in Atlanta.

The children were now parents, usually having become so at an early age. In fact, every single girl Heath had studied in the original community had had a child during her teenage years. The new mothers did not have baby-talk games with their children and did not label things in the environment for them. They did not ask their children to tell them about their day. When the child did talk, the mother sometimes ridiculed any display of knowledge.

In many cases, the children became the charge of the teenager's mother. The teenager returned to high school and the social life there. One of the teenagers Heath studied, who was on welfare most of the time, rarely engaged in much interaction with her children or enlisted them in talk of any kind, and when she did, the interaction typically lasted less than a minute. Her world was much less rich linguistically and socially than the extended family she had been born into in North Carolina. Her language experience was limited to passively watching TV or reading movie or TV magazines. The people she associated with were the largely transient women in the project where she lived. Talk of the future for these women was of the immediate future only—how to get documents to the welfare office or how day-care rules might change or how she could get the children's father to start sending money again.

It is clear from Heath's report that for the children of at least some of the young parents of the late 1980s, their cognitive, social, and emotional lives were far less rich than their parents' had been when growing up.

More systematic studies of the home life of a representative sample of black families in the 1980s support Heath's view that the home environments of blacks in that era and now can be intellectually impoverished and emotionally harsh. Meredith Phillips, Jeanne Brooks-Gunn, and their colleagues have looked in detail at the results of studies measuring aspects of the home environment of blacks and whites. Their analysis was based on two data sets. One came from a study called Children of the National Longitudinal Survey of Youth, or CNLSY. The study began in 1986 and covered more than six thousand children born to people who were between the ages of fourteen and twenty-two in 1979. A wide range of demographic and home variables was studied. The second data set came from the massive Infant Health and Development Program (IHDP), which examined children born at eight different hospitals who had birth weights lower than 2,500 grams (about five and a half pounds)—a figure that is considered to place them at risk for low IQ as well as a range of physical health problems. In the next chapter I discuss the effects of an extremely ambitious intervention program on the subsequent IQ and academic achievement of these children. For the time being I consider only the 315 black and white children in the control group in IHDP whose birth weight was between 2,000 grams and 2,500 grams.

The measures that Phillips and her colleagues examined are contained in the so-called HOME scale (Home Observation for Measurement of the Environment). Scores on this measure are based on an interviewer's observations in the home and questions asked of the mother. Things studied include "learning experiences outside the home (trips to museums, visits to friends, trips to the grocery store), literary experiences within the home (child has more than 10 books, mother reads to child, family member reads newspaper, family receives magazine), cognitively stimulating activities within the home (materials that improve learning of skills such as recognition of letters, numbers, colors, shapes, sizes), punishment (whether child was spanked during the home visit; maternal disciplinary style), maternal warmth (mother kissed, caressed, or hugged the child during the visit; mother praised the child's accomplishments during the visit), and the physical environment (whether the home is reasonably clean and uncluttered; whether the child's play environment is safe)."

Differences between black and white homes in the two studies were marked and ranged as high as three-fifths of a standard deviation on some measures. Within the black sample, scores on the HOME scale were closely associated with scores on cognitive variables. For the CNLSY study, the vocabulary score for five- and six-year-olds was equivalent to an additional 4 IQ points when the mother read to them daily as opposed to not at all. For the IHDP study, IQ scores were 9 points higher when a family scored 1 standard deviation above the mean on all the HOME measures.

Hart and Risley, in their study of Kansas families, which I described in Chapter 5, found huge differences across groups in how parents treated their children in terms of warmth versus punishment. Recall that the children of professionals received six encouragements per reprimand and the children of working-class parents received two encouragements per reprimand. The children of black welfare parents received two reprimands per encouragement. By the time the child of white professionals is three years old, she has heard 500,000 encouragements and 80,000 discouragements. The three-year-old black child whose mother is on welfare has heard about 75,000 encouragements and 200,000 reprimands.

There is every reason to believe that such differences could be important for cognitive development. And we know that the discouraging aspects of black parenting are characteristic even of middle-class black parents to a degree, or at least they were in the 1980s. Recall the study by Elsie Moore comparing black and interracial children raised by middle-class black or middle-class white parents. The IQs of black and interracial children raised by white adoptive parents were 13 points higher than those of black and interracial children raised by black adoptive parents. We have no idea how much of this difference was due to the home environment and how much was due to the school and neighborhood environment. We do know, however, that the home environment of children raised in the black homes was not as likely to promote cognitive growth as much as that of children raised in white homes.

Moore's researchers went into each home and had children work on a block design task, in which the child looked at a picture and tried to duplicate it using blocks, while the mother was present. White mothers were substantially more encouraging and less disapproving than black mothers. When the child was having trouble with a design, the white adoptive mothers tended to diffuse tension by joking, grinning, and laughing. The black adoptive mothers were more likely to frown and scowl. The white adoptive mothers encouraged the problem-solving efforts of their children ("Gee, that's an interesting idea" or "You're good at this"). Black mothers were more likely to express disapproval ("You know that doesn't look right" or "You could do better than this if you really tried"). The white mothers also provided help to their children in ways more likely to promote learning, such as suggestions that the child could use to explore for himself how to work on the designs ("Why don't you work on it one section at a time?"). Black mothers were likely to give specific instructions, which left no opportunity for the child to discover how to complete the task himself ("That will work, but you have to turn it around like this," and then shows the child the appropriate maneuver). The white mothers' attitude was essentially, "It's okay to be wrong if you're trying." The black mothers were more likely to display exasperation and lack of confidence.

I want to stress two points concerning this evidence about the behavior of black and white mothers. First, we do not know how much it was this behavior that held back the child's intellectual development and how much it was other environmental factors such as neighborhood, peers, or schools. Second, it is unlikely that such marked differences would be found between black and white middle-class mothers today. The study is now almost twenty-five years old, and second-generation middle-class parents likely behave differently from first-generation middle-class parents. Certainly we would expect the second-generation middle-class mothers to behave in ways that would more likely promote exploration and growth.

The conclusions of the previous chapter and the present one dovetail. Genes account for none of the difference in IQ between blacks and whites; measurable environmental factors plausibly account for all of it. Lower-class blacks and lower-class whites suffer from many of the same disadvantages, but blacks, especially the black working class and underclass, suffer from racial prejudice that blocks their occupational trajectories. Some aspects of black culture—at every social-class level—are less likely to promote cognitive performance compared with white culture. The neighborhoods and schools available to all but middle-class blacks both compound the deleterious effects of that culture and make movement out of it difficult. And even for middle-class blacks, there are adolescent male subcultures that are anti-achievement. These subcultures encourage the belief that athletic skill, a talent for entertainment, and street smarts can substitute for academic skills.

I do not doubt that in the normal course of events, slow progress will occur in the socioeconomic and intellectual aspects of black life. Crime rates and drug usage rates have been coming down steadily over the last few decades (though there has been an increase in violent crimes since 2005). Moreover, black entry into the middle class, and improvement in black IQ and academic achievement scores, continue to increase.

In the next chapter you will see whether anything can be done to hasten the movement of poor blacks into the ranks of the working class, and of working-class blacks into the ranks of the middle class.

CHAPTER SEVEN

Mind the Gap

> *Compensatory education has been tried, and it apparently has failed.* —Arthur Jensen (1969)

> *There is no evidence that school reform can substantially reduce the extent of cognitive inequality as measured by [ability] tests.* —Christopher Jencks and others (1972)

> *There is no reason to believe that raising intelligence significantly and permanently is a current policy option, no matter how much money we are willing to spend.*
>
> —Charles Murray (2007)

IN 2002 THE U.S. CONGRESS passed the No Child Left Behind Act, which mandated that American schools eliminate the gap between the social classes and between minority groups and whites by 2014. I don't know if most members of Congress actually believed that such accomplishments are possible. But if so, they are deeply ignorant of the forces that operate to produce high academic achievement.

Intellectual capital is the result of stimulation and support for exploration and achievement in the home, the neighborhood, and the schools. To think that this can be changed by mandate—operating only through the schools—is preposterous. Moreover, the schools attended by minorities and the poor are wanting in ways that cannot be drastically improved overnight. The problems include quality of teachers willing to work in these less rewarding schools, the caliber of school management, the disruptiveness produced by high levels of student turnover, and the nature of the schools' clientele, whose homes and neighborhoods make it unlikely that they will be encouraged toward high academic achievement.

It should be clear from the previous chapter that there is no theoretical limit on the degree to which the achievement gap between blacks and whites can ultimately be closed. Though there is far less evidence on the native intellectual ability of the extremely broad and diverse group of cultures labeled as "Hispanic," I see no reason why the gap cannot ultimately be bridged there as well.

On the other hand, it should be clear that unlike the black/white and Hispanic/white gaps in achievement and IQ, the social-class gap is never going to be closed. This is true, if for no other reason, because the well-off are always going to find ways to get a better education for their children and are always going to find ways to be ahead in terms of parenting skills and are always going to be able to provide superior neighborhood environments. In addition, there is always going to be at least some difference in the gene pools of the lower class and the middle class. Recall from Chapter 1 that within a given family the sibling with a substantially higher IQ achieves much higher socioeconomic status (SES) than less favored brothers and sisters. And since the higher IQ is attained in part by virtue of a better luck of the draw from the gene pool of the parents, higher SES is always going to be in part a result of better genes for intelligence. So higher-SES people are going to pass along better prospects for intelligence to their offspring by virtue of having, on average, better genes and by offering better environmental advantages to their offspring.

But these considerations should not be cause for pessimism about the degree to which the intellectual lot of lower-SES people can be improved. Recall from Chapter 2 (on heredity) that the effect of an upper-middle class upbringing on children born to lower-SES parents is to raise the IQ by 12 to 18 points. The theoretical ceiling for improvement of lower-SES intellectual capital is very high indeed.

But how much improvement can we realistically hope to produce for lower-SES individuals and for currently disadvantaged minorities?

Early Childhood Education

When I tell people that I am writing a book on the modifiability of intelligence, they sometimes tell me, so as to encourage me to avoid wasting my time, that Head Start does not work. For many people their belief about one particular program settles the matter of whether intelligence can be manipulated.

Head Start is a compensatory program initially aimed primarily at improving the health and welfare of poor children three and four years old. Some of the founders hoped that it would also lead to improvements in the children's intelligence, school performance, and subsequent success in life. Head Start sessions run for a half-day five days a week, typically for thirty-four weeks, but only a small portion of each session specifically targets cognitive concerns.

Is Head Start a failure? It depends on your perspective. With respect to physical health, Head Start has been a resounding success. It results in mortality rates for children that are 33 to 75 percent lower than for comparable children not in the program. It has driven mortality rates down, in fact, to a level not substantially different from those for children in general.

In earlier days, Head Start was associated with a gain of about .35 SD on cognitive tests, or about 5 IQ points, when the child finished the program at age five, and more recent studies showed that there is still a .10 to .20 SD effect on some IQ and achievement variables at age six or seven. These fade into nothingness by late elementary school. Recent reports show lower effect sizes at age five—on the order of .25 SD. But it is difficult these days to find true control groups for pre-kindergarten interventions because the majority of even poor minority children receive some kind of pre-kindergarten care. As a result, effects are compared, not to no-treatment controls, but to control groups in which typically half of the children have some day care.

There are shockingly few evaluations on the long-term academic effects of Head Start. What little there is indicates a slight effect of Head Start on completion of high school, which is about

2 to 5 percent greater than for controls, and a small effect as well on likelihood of attendance in college, which is about 3 to 6 percent greater than for controls. The cost of Head Start is about $7,000 per child, so whether the program is worth the gain from an intellectual and academic standpoint is an open question.

Early Head Start, which begins at birth and continues to age three, has proved to be no more successful than Head Start in improving educational outcomes. Services include child development, child care, home visiting, parenting education, and family support services. Individual programs have been given substantial latitude about which services to emphasize. Effect sizes on a range of variables from the purely cognitive to the emotional and social are in the range of .10 to .30 SD—slightly higher for minority children than for white children. Even the best versions of the program produced IQ gains of less than 4 points in the short term (though vocabulary scores increased by .40 SD). The program is expensive, and it is not clear whether long-term gains (which apparently will not be measured) would justify the cost.

But there are more ambitious programs than Head Start, and some of them have much bigger and more lasting results. A review of about a dozen of the better small-scale preschool and kindergarten programs focusing on black children showed that they lead to big gains in IQ—of as much as .70 SD or even more at age five. They also result in significant achievement gains in the first few grades of elementary school, but these gains generally fade, often completely.

Fading is to be expected if high-quality environments are not maintained. Only if children's brains are like clay would we expect them to remain in good shape years after they were formed. If children's brains are like muscles, however, then we would expect exercise, in the form of stimulating environments and activities, to be necessary to maintain good performance. I favor the muscle view, and so do the data.

As it turns out, several early childhood education programs actually do produce large immediate gains in IQ, as well as long-term gains in IQ or academic achievement, or both. Let's look at three of the more effective programs that randomly assigned children to treated versus untreated conditions and that followed the children until late adolescence or adulthood.

The Perry Preschool Program was carried out in Ypsilanti, Michigan, between 1962 and 1967 by Lawrence Schweinhart and David Weikart. It was administered to fifty-eight black children living in poverty whose mothers had IQs between 75 and 85 on the Stanford-Binet IQ test. Children entered the program at age four in the first year of operation and at age three in the remaining four years. With the exception of the first group, children spent two years in the program.

The Perry treatment consisted of daily morning sessions in a classroom lasting two and a half hours for thirty weeks each year and focused on activities that would foster cognitive growth and social development. The average child-teacher ratio was 6 to 1, which is extremely low, and staff were very well trained in early childhood development and education. Once a week, the teacher would also visit each child's home for ninety minutes to encourage the mother to become involved in the educational process. The fifty-eight children comprised the treatment group, and sixty-five were in a control group. When children completed the program, they entered school in the disadvantaged neighborhoods where they lived.

The mean IQ of the children in the control group at the end of the program, at age five, was 83. The mean IQ of the treatment group was 95. The IQ of the treatment children dropped progressively over the years of grade school until it was the same at age ten as the control mean, 85.

The IQ results are disappointing, though scarcely surprising given the circumstances of home, neighborhood, and school after termination of the program. What is surprising is that the academic gains and subsequent economic and social gains were enormous. These are summarized in Figure 7.1A on page 128. About a third of the control-group children were assigned to special education classes at some point, versus 13 percent of the treatment group. At age fourteen, only 14 percent of the control group tested above the 10th percentile on the California Achievement Test, compared with almost half of the treatment group. Effect sizes for reading, math, and language achievement scores ranged from 0.50 to 0.75 SD. Forty-three percent of the control group managed to graduate from high school, compared with 65 percent of the treatment group. High school grades were .57 SD higher for the treatment group than for the control group. At age twenty-seven, only about 6 percent of the control group earned as much as $2,000 per month versus 28 percent of the treatment group. Eleven percent of the control group owned their own home, whereas 33 percent of the treatment group owned their own home. About 20 percent of the control group had managed never to be on welfare, compared with 40 percent of the treatment group. Eight percent of the control-group women and 40 percent of the treatment-group women were married. By the age of forty, 55 percent of the controls had five or more arrests; 36 percent of the treatment group had five or more arrests. These results have tremendous social and economic implications.

The Perry program is by no means unique in showing fading IQ gains for intervention groups combined with big achievement advantages later in life. These advantages include lower percentage retained in grade, lower percentage in special education, and higher percentage graduating from high school. The fact that gains in academic and life achievement can be so great even though IQ gains had completely dissipated has led many to suspect that the achievement gains are attained not through increased intelligence per se but rather through temperamental or motivational changes that resulted from the intervention and that persisted even when the IQ gains were no longer supported by the environment.

An intervention even more ambitious than the Perry Program, which lasted much longer, was initiated by researchers in Milwaukee. One particular area of the city, which had 3 percent of the population, accounted for 33 percent of the mentally retarded children in the district. The investigators decided to concentrate their resources on that section of the city. All of the children recruited for the study were African Americans at high risk for mental retardation because their mothers were poor and had IQs of 75 or less. The children were randomly assigned either to a control group (eighteen children) or to an intervention group (seventeen children), which was an intensive day-care program lasting from the time the children were less than six months old until they enrolled in first grade.

The intention of the Milwaukee Project was to give children the equivalent of a middle-class environment. The program focused on developing their language skills and cognitive capacities. Paraprofessionals frequently interacted with the children in an enjoyable way, using the best developmental programs and educational toys known at the time. Sessions lasted seven hours each day, for five days each week. The program provided the children with good food and high-quality medical and dental care and offered training in homemaking and child care to the mothers. The treatment group was compared not only to the control group but also to a low-risk comparison group of children born to mothers of average to above-average intelligence (108 on the Wechsler Adult Intelligence Scale). At the age of thirty months, the control-group children had an average IQ of 94 on the Stanford-Binet test, compared to 124 for the treatment-group children. The treatment group actually scored higher than children born to mothers with average or above-average IQs. Those children scored 113. By age five, the control group scored 83 on the Wechsler Preschool and Primary Scale of Intelligence, compared to 110 for the treatment group, which was still higher than the IQ score of 101 for the low-risk comparison group.

At termination of the program at age seven, the average treatment-group IQ was still 22 points higher than the average for the control group. Note that this gives an even higher upper bound for the difference between typical lower-SES rearing strategies and superior strategies. Perhaps this bigger difference occurred because the Milwaukee children, unlike the Perry children, went to reasonably high-quality schools—all rated at or above the national average for scores on achievement tests. In addition, children in the treated group maintained substantial IQ gains. Nine years after the program was over, when the children were adolescents, the control-group children scored 91 on the Wechsler Intelligence Scale for Children and the treatment-group children scored 101—on par with the 97 IQ of the comparison group born to mothers with average or higher IQs.

Achievement scores for grades 1 through 4 for the treatment group were higher than those for the control group, and the difference was great in standard deviation terms—around .75. The number of children was low, however, and there is a 10 percent probability that a difference that big could have been obtained by chance (though there is only one chance in twenty for a reliable difference specifically in favor of the intervention group, as opposed to a difference in favor of either group).

A yet more intensive intervention was attempted by Craig Ramey, Frances Campbell, and their colleagues in a project called the Abecedarian Program. Almost all of the 111 children involved in the program, who were born between 1972 and 1977, were African American. The children were considered as being at high risk of retardation in light of their mother's IQ, which averaged 85, and mother's education, which averaged ten years, in addition to other risk factors such as low family income, absence of father, poor social or family support for the mother, poor academic performance of siblings, employment of parents in unskilled labor, and reliance on public agencies for support. Abecedarian was a full-day intervention that began before children were six months old and lasted all year-round. The infant-teacher ratio was 3 to 1 at first and increased to a child-teacher ratio of 6 to 1 as the program progressed. The intervention program continued up to kindergarten age. Data were collected on the participants regularly up until age twenty-one.

The program included four groups of about twenty-five children each. One group was assigned to the infant-kindergarten intervention and also to a school-age intervention. The latter intervention provided, for the first three elementary school grades, a home-school teacher who met with the parents and showed them how to supplement the educational activities at home. Parents were encouraged to work with the child for at least fifteen minutes a day. This home-school teacher was also a link between the school teachers and the family. She met with teachers and families once every two weeks. She also helped the family with finding jobs, dealing with social service agencies, and taking the children to doctor appointments. One group of children was assigned only to the preschool intervention, and one group only to the school-age intervention. The fourth group was assigned to neither intervention.

At age three, children who had no preschool had an average IQ of 84; children who received the preschool intervention had an average IQ of 101. At termination of the intervention, control children had an average IQ of 94; preschool intervention children, an average of 101. Then, instead of attending poor-quality inner-city schools, all children went to schools where most of the children were reasonably well-off white kids. At age twelve, only 13 percent of children exposed to the intervention had IQs of less than 85, compared to 44 percent of the control children. Even at age twenty-one, those who had the preschool intervention had an average IQ 4.5 points higher than the average for those who had been in the control group. The children whose mothers had the lowest IQs (less than 70) benefited the most from the program. There is no evidence that the school-age treatment added anything in the way of IQ to the preschool treatment, nor did the school-age treatment by itself accomplish much. This study is one of many prompting pessimism about the effects of home visits—unless they are very ambitious in their scope.

By the time the subjects in the study reached the age of twenty-one, it was clear that the Abecedarian preschool intervention had had a major influence on many educational outcomes, summarized in Figure 7.1.B. Almost half of those in the control group had been assigned to special education classes at some point in their educational career, versus fewer than a fourth of those in the intervention

A

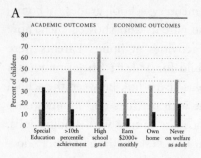

B

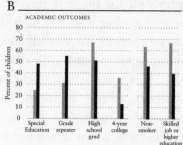

Figure 7.1. Academic, economic, and social outcomes for the Perry Preschool and Abecedarian programs. (A) Data from the Perry Program collected when the individuals were twenty-seven years old. >10th percentile achievement, children who scored above the lowest 10 percent on the California Achievement Test (1970) at age fourteen; HS Grad., percentage of children who graduated from high school on time. (B) Data from the Abecedarian Program collected when the individuals were twenty-one years old (Carolina Abecedarian Project and the Carolina Approach to Responsive Education, 1972–92). Light bars, intervention group; dark bars, control group. From Knudsen, Heckman, Cameron, and Shonkoff (2006).

group. Over half of the control group had repeated a grade, whereas 30 percent of the intervention group repeated a grade. At age fifteen, the reading scores of the intervention group were 1.40 SDs higher than those of the control group, and the math scores were .86 SD higher than those of the control group. By age twenty-one, the intervention group was two grade-years ahead of the control group in reading scores and more than a year ahead in mathematics scores. Half of the control group had graduated from high school, versus two-thirds of the intervention group, and 12 percent of the control group had attended a four-year college, versus a third of the intervention group. At age twenty-one, fewer than 40 percent of the control group were either in a skilled job or in higher education, compared with two-thirds of the intervention group.

The Abecedarian effects reported are probably underestimates of the actual effects because the control-group children received pre-kindergarten care of some kind.

There have been several replications of at least part of the Abecedarian intervention. One is important because it refutes the contention by Herrnstein and Murray that the results of the Abecedarian study are suspect since substantial IQ differences were found at age one. Herrnstein and Murray believed such differences could not be due to the program at such an early age and therefore were indicative not of program effectiveness but of a failure of random assignment to produce equivalence of control and intervention groups at the outset. Project Care, using methods essentially like those of Abecedarian, started with children assigned to intervention versus control groups who had identical scores on the Bayley Mental Development Index at six months. By age one the scores for the two groups differed by 11 points. So Abecedarian has effects on the IQ of even very young children.

Another replication of Abecedarian is particularly important because it shows that the program can be used to boost the IQ of children at risk of mental retardation owing to prematurity and low birth weight—2,001 to 2,500 grams. Two-thirds of the infants in the study were black or Hispanic. About 40 percent of the mothers had not graduated from high school, and only about 13 percent had completed college. The program was different from Abecedarian in that rather than starting shortly after birth, it began at age one, and continued only until age three. At the end of the intervention at age three, IQs of the intervention group averaged 9.2 to 12.5 points higher than control IQs, depending on the test used.

Two years after the end of the intervention, IQs of the intervention group were 2.5 to 5.4 points higher, depending on the test used (the lower estimate was not statistically significant). At age eight, there were still detectable effects of the intervention, with intervention-group IQs ranging from 3.6 to 5.4 points higher, depending on the test used. Even at age eighteen, there was a detectable difference in IQ of 3.8 to 5.3 points between youth who had been in the intervention program and those who had not. Achievement differences were not impressive at any point.

IQ gains for the intervention group were much larger for the children who stayed in the program the longest. But this kind of analysis—showing that participants who stayed longest benefited most—invites a self-selection bias, which was not completely compensated for by control procedures used by the investigators. And it should be noted that treatment effects were not significant beyond age three for infants who weighed extremely little at birth (less than 2,001 grams).

A particularly important fact is that early childhood intervention programs benefit black and Hispanic children more than white children (who already have more of the advantages conferred by the interventions, on average), and benefit poor children more than middle-class children.

In short, early childhood intervention for disadvantaged and minority children works—when it is strenuous and well conducted. Many different programs get high gains in IQ by the time they end. These gains generally fade over the course of elementary school, but there is some evidence that this is less true if children are placed in high-quality elementary schools. Much more important are the achievement gains that are possible: lower percentage of children assigned to special education, less grade repetition, higher achievement on standardized tests, better rates of high school completion and college attendance, less delinquency, higher incomes, and less dependence on welfare. And these changes can be very large.

There has not been much research on the effectiveness of home-visitation programs having the intent of improving parenting practices. Some reasonably ambitious parenting interventions led to improvements in mothers' behaviors as well as children's emotional and cognitive behaviors. The key to success appears to be coaching parents on specific behaviors.

One particularly effective intervention indicates that such programs might have great value. This was conducted by Susan Landry and her coworkers. They went into the homes of mostly lower-income, black and Hispanic mothers of one- to two-year-old infants and provided ten to twenty 1½-hour sessions showing the mothers beneficial patterns of response to their infants. They worked on teaching mothers how to interpret the intent of their child's positive and negative signals, how to respond to the child's behavior in warm and sensitive ways even when the child denies the mother's requests, how to attend to the child's focus of attention and maintain and build the child's interest, when to introduce interactions and games, and how to use vocabulary-rich language, labeling objects and actions. The effect on mothers' behavior was marked for many dimensions, including warm and sensitive behavior, responding appropriately to child's intents, maintaining the child's interest in activities, and verbally encouraging the child. Effect sizes for such outcomes ranged as high as 1 SD.

The effect on the children's behavior was also substantial. Children became more cooperative, were more engaged when interacting with their mothers, used words more frequently, were better able to coordinate their words with what they were attending to, and scored higher on a picture vocabulary test. These effects ranged as high as .70 SD.

We do not know yet what the long-term effects of parental interventions are, but there seems to be good reason to be hopeful. The benefit-cost ratio of such programs could be very high.

School-Age Interventions

How about interventions for school-age children? What can be done for children who have not had powerful preschool interventions, or could be done to sustain the gains experienced by children who have been in center-based day-care programs?

I'll start with some bad news—and unfortunately there is a lot of it. In Chapter 4 (on improving schools) I discussed some of the efforts made to improve the academic performance of children in general. Sheer amount of money spent doesn't do a lot of good for students in general, so we would not expect money by itself to narrow the achievement gap between lower- and upper-SES students or between minority and white students. Vouchers for attending private schools have been given out to poor and minority children, but there is not much evidence that they are the answer. The same is largely true for charter schools. Some might do a better job than the public schools—in fact, you will read about one later—but charter schools in general seem to be little better than the public schools even after they have been in operation for a while, and they may actually be worse in their early years.

Are there some regular public schools that do a particularly good job with disadvantaged minority populations? Two different reports maintain that at least some schools do a superb job with their underprivileged, minority clientele. One report is by a conservative institution, and one is by a liberal institution.

The Heritage Foundation, a conservative institution, claims to have found twenty-one high-poverty schools whose students have achievement levels above the national norms. These schools are held up as mavericks that escape from low performance by dint of having visionary principals who are willing to buck the "cult of public education," fire bad teachers, and teach the basics instead of progressive nonsense.

Richard Rothstein, the former national education columnist for the *New York Times,* debunked these claims. According to him, only six of the twenty-one schools were fully nonselective neighborhood schools. The rest were (a) magnet schools; (b) schools where many of the parents were "impoverished" graduate students; (c) schools that did indeed produce high scores early on by concentrating on the basics such as phonics, but scores declined precipitously in later grades precisely because the emphasis on the basics precluded learning reasoning and interpretation skills necessary for success in the upper grades; or (d) schools where parents had to apply to get their children accepted, thereby introducing a potentially heavy self-selection bias.

The Education Trust, a liberal organization, claims to have uncovered 1,320 schools in which at least half the students were both poor and minority and whose test scores were in the top third for their states. The claims about these schools do not hold up either, according to Rothstein. The 1,320 schools did in fact have high scores, but in only one grade, in only one subject, and for only one year. These accomplishments were mostly statistical flukes.

Rothstein gives an even more bracing back of the hand to yet another claim, this one by Douglas Reeves, who says he has identified a group of "90/90/90" schools in Milwaukee where 90 percent of the students are poor, 90 percent are minority, and 90 percent meet "high academic standards." Those standards turn out to be only basic, nonproficiency level scores as defined by the state of Wisconsin.

How about claims that if you put poor minority kids into a school with children whose families are well off, their academic achievement can soar? The *New York Times* announced on its front page in 2005 (and then repeated in another front-page story in 2007, and yet again in a magazine story in 2008) that such effects had been achieved in Wake County in Raleigh, North Carolina. Under the headline "As Test Scores Jump, Raleigh Credits Integration by Income," the *Times* said this:

> Over the last decade, black and Hispanic students here in Wake County have made such dramatic strides in standardized reading and math tests that it has caught the attention of education experts around the country.
>
> The main reason for the students' dramatic improvement, say officials and parents in the county, which includes Raleigh and its sprawling suburbs, is that the district has made a concerted effort to integrate the schools economically.

The article goes on to state that since 2000, officials used income to guide assignment to schools, with the intention of holding the proportion of low-income students to less than 40 percent. The result?

In Wake County, only 40 percent of black students in grades three through eight scored at grade level on state tests a decade ago. Last spring, 80 percent did. Hispanic students have made similar strides. Overall, 91 percent of students in those grades scored at grade level in the spring, up from 79 percent 10 years ago.

Unfortunately, there is no evidence that schools in Raleigh do a better job for minority students than schools statewide. The black/white gap was actually slightly less for students in the state as a whole in 2004/05 than it was in Wake County. Moreover, the apparent gains in Wake County by minority students were almost surely due to an easing of statewide standards for proficiency. By the new standards 95 percent of whites were proficient, which indicates that the standard was so lenient it is impossible to get a read on the actual size of the gap.

The data comparing Wake County with the state as a whole are encouraging in one respect, though. They indicate that integration of low-income children with higher-income children may hurt the academic achievement of neither—and may result in social gains for both. But we would have to know much more—for example, the SES level of blacks and whites both statewide and for Wake County—before we could reach that conclusion. There is good evidence, incidentally, that black and Hispanic children learn more in integrated classrooms than in majority black or Hispanic classrooms.

How about efforts to improve schools by instituting top-to-bottom changes in administration and curriculum? We know from Chapter 4 that there are a lot of such whole-school interventions, in which entire curricula and educational strategies are set in place in a school. But those interventions tend not to produce very impressive results for students in general, so we have to be skeptical as to whether they would be very successful in reducing the achievement gap.

Teacher certification and higher academic degrees do not improve the achievement of anyone very much either. Experience in teaching counts, though, and there is a possibility that experience matters more for lower-class than middle-class kids or more for minority kids than for whites. Teacher quality also makes a big difference for students' achievement scores. Again, there is the possibility that teacher quality matters more for poor and minority kids. We certainly know that Miss A, heralded in Chapter 4, made a huge difference for the poor kids in her first-grade class. And we know that the performance of poor kids and of kids who had been disruptive in kindergarten is greatly affected by the quality of instructional support and emotional care that they get in first grade.

We also know that smaller class sizes result in better performance on achievement tests, and that the effects are bigger for black kids than for white kids—.33 SD versus .25 SD—and bigger for poor kids than for middle-class kids.

There are some big success stories in K–12 education though. Two educational programs for poor and minority children—one for math and one for reading—have been shown to be quite effective.

The math training program is called Project SEED. It is an enrichment program that hires and trains mathematicians, scientists, and engineers to teach poor minority students. The program teacher uses Socratic questioning to introduce abstract mathematical concepts and students become active participants in the lessons—through the use of dialogue, debates, and choral response. The regular teacher sits in on all sessions. Project SEED is not a replacement for the regular mathematics curriculum but an add-on. One study of Project SEED's effectiveness, carried out in Dallas, compared the California Achievement Test (CAT) scores of 244 fourth-grade students in SEED classrooms with 244 students in the same schools who did not have SEED instruction and 244 students who were not in SEED schools but were deemed comparable to them. The SEED students outscored the comparison-group students by .37 SD—a very significant benefit. But SEED students outperformed the non-SEED students in their own school by only .19 SD. Arguments as to which is the more appropriate control can be made both ways, and it makes a good deal of difference which side one comes down on. A gain of .37 SD is surely worth the cost, which was not great; a gain of .19 SD may not be.

Several reading programs geared to minority and poor children have been developed, of which one of the most promising is Reading Recovery (Descubriendo La Lectura in its Spanish version). Reading Recovery is a tutoring program for low-performing first-graders that was developed by Ohio State University researchers. Tutors provide a daily one-on-one 30-minute lesson for twelve to twenty weeks. Children read stories they already know verbally, read a story they read the day before, and write stories. The Ohio State group conducted randomized studies to evaluate the program. They found effect sizes ranging from .57 to .72 SD for most indicators of reading proficiency. The effect sizes faded over time, though they were still detectable at about .20 SD in the third grade. One independent study evaluating the Spanish-language program demonstrated effect sizes ranging from about 1.00 to 1.70 SDs. This finding would need to be duplicated before we could place much confidence in it.

There is at least one extremely important exception to the rule that whole-school interventions, and charter schools, have only modest effects on student achievement: the Knowledge Is Power Program, or KIPP. Michael Feinberg and David Levin, two Houston-area elementary-school teachers with four years of teaching experience between them, founded this extraordinarily ambitious educational project in 1994. They designed it around the needs of poor children, especially minority children. They freely admit that they invented the program as they went along. But they had help from a much more experienced mentor—Harriett Ball—a teacher who had grown up in the racially segregated neighborhoods of Houston and who was known for having classes with students who were well behaved and high achieving.

Feinberg and Levin developed a type of school, initially mostly for middle-school children, based on 7:30 a.m. to 5:00 p.m. school days (yes, 9½ hours), mandatory attendance for three additional weeks in the summer, a half-day of Saturday classes biweekly, visits to students' homes, an insistence on kindness and good behavior, the principal having the power to hire and fire teachers, cooperation among teachers, and a system of rewards and penalties for behavior and academic accomplishment. The huge amount of extra contact time allows for the KIPP students to get exposure to what upper-middle-class students get through their homes and expensive public or private schools—sports, museums, dance, art, musical instruments, theater, and photography. The first two schools—in Houston and the Bronx—scored highest on achievement tests of any schools in their areas. Since 2001 KIPP has begun to expand with financial support from Doris and Don Fisher, the founders of Gap stores. Most KIPP schools are charter schools that are essentially franchises of the KIPP Foundation. Two other similar programs, called Achievement First and North Star, have been less well researched.

KIPP's students are economically disadvantaged as a group. More than 80 percent are eligible for federal free or subsidized lunches. Most are African American or Hispanic. KIPP maintains that "while the average fifth-grader enters KIPP in the bottom third of test-takers nationwide (28th percentile), the average KIPP eighth-grader outperforms nearly three out of four test-takers nationwide (74th percentile) on norm-referenced reading and math assessments."

But some KIPP schools have posted losses for their students (to KIPP's credit, by its own admission), and KIPP's claims for successes have been based mostly on its own teachers' tests and not on independent research. However, SRI International conducted an independent study of KIPP schools in the San Francisco Bay Area, in the early years of this century, and I report on it at length here.

In the five Bay Area KIPP schools, 72 percent of the children were economically disadvantaged and 75 percent were African American or Latino. Each school in the study was compared to two comparison schools that were similar socioeconomically and with respect to minority composition. The KIPP schools began in

2002/03 with an entering class of fifth-graders and added a class each year.

Principals study the KIPP model for a year before they get their own schools. But the KIPP model does not prescribe a particular instructional approach or curriculum, so teachers do not have to learn a new system required by the model, in contrast to most whole-school reforms. At the time of the study, about half the Bay Area KIPP teachers came from the Teach for America program, and their median teaching experience before joining KIPP was two years.

It is possible for children to fail in KIPP schools. Principals believe that the option to hold students back is essential because the school "couldn't let other kids see that kids who did nothing could move ahead." It is impressed on staff, students, and parents that certain behaviors are demanded of them. The KIPP credo is "If there is a problem, we look for a solution. If there is a better way, we find it. If we need help, we ask. If a teammate needs help, we give it." Some sample KIPP slogans are "Work hard, be nice," "All of us will learn," and "KIPPsters do the right thing when no one is watching."

Students earn "paychecks" each week, based on a point system that adds on to their value or subtracts from it depending on behavior and academic performance. The paycheck can be used to buy items from the KIPP store, including snacks and school supplies, and to buy attendance on field trips. Students get publicly "benched" for bad behavior and failure to do work. At one school, students have to go three days in a row without getting deductions from their paychecks in order to get off the bench. Teachers insisted to SRI researchers that all student actions have consequences. But teachers also maintained that good behavior was not due to fear.

"We've never had a kid talk back to a teacher, and we've never had kids fight. I don't attribute this to the discipline system. It's from setting expectations from the start. The smallest detail was called out . . . It's because kids believe that this is an extraordinary place, and we've taught them that. I don't think they don't tease because they are afraid of the [bench]. It's just something that they would not do at KIPP. This is the one school they've been to where there's no teasing. They feel safe, and they are learning more."

Another teacher said, "At this school, it's okay to be smart, and that's something that is lacking at most inner-city public schools . . . In [my former district], I worked with traditionally under-served kids and found the schools to be an inadequate solution to kids' needs . . . I visited a KIPP school [nearby], and the school was like an oasis."

Comments of students show that they are well aware of the difference between the KIPP school and others they have attended. "Everyone is committed to learning." "The other school was not challenging enough, and I've found it has been here." "Now I realize you have to work to get to college." In all the schools, students said there were fewer fights, if in fact there were any at all, than in previous schools they had attended: "At this school, something holds me back from fighting."

Of course the demands on teachers are enormous. They must be in school from 7:15 a.m. to 5:15 p.m. and are usually there longer. In addition to planning curriculum, they provide instruction, monitor study halls, lead enrichment activities like trips to zoos and museums, tutor, and call parents at night to discuss their children's performance. They also work several weeks during the summer and every other Saturday. Not surprisingly, they fight emotional exhaustion. And in general, most KIPP teachers say they will be able to stay for only a few years.

What educational results does all this produce? Children in KIPP schools in the Bay Area succeed at levels well beyond what would be expected considering their demographic composition. The children were given the reading, language arts, and math Stanford Achievement Test (SAT 10), which is normed nationally, in the fall and spring. For fifth-graders, fall scores indicate where students are at the outset of the KIPP experience. The improvement from fall to spring was calculated in terms of the percentage of students scoring at or above the 50th percentile for national norms at each point. In the fall, for fifth-grade language arts, the proportion of students scoring at that level was 25 percent (on average for all classes at the four schools that gave the test), slightly more than we would expect given the demographic characteristics of the schools. In the spring, it was 44 percent, substantially higher than we would expect given the demographics. In the fall, for fifth-grade math, the proportion was 37 percent; in the spring it was 65 percent. Gains for sixth-graders were also marked. These increases are extremely large. They indicate that after just one year at a KIPP school, disadvantaged and largely minority children were scoring close to or above the national average on standardized tests. Since scores were not very impressive at the beginning of the year for those same children, we can rule out the possibility that self-selection is entirely responsible for the outcome in the spring.

In the spring, all children (at all five schools) were also given the California Achievement Test, which is required by law. For English Language Arts, 43 percent of KIPP fifth-graders scored at the level of proficient or above, versus 19 percent of the fifth-graders at comparison schools chosen for their demographic similarity to the KIPP schools. For Mathematics, 55 percent of KIPP fifth-grade students scored at the level of proficient or above, versus 20 percent of comparison-school students. Results on the CAT were entirely comparable for the sixth grade. Again, these results are extremely striking. (Baseline, beginning-of-the-year CAT scores were comparable for KIPP and the comparison schools.)

I have violated my own implicit research standards in reporting the SRI study. Students were not randomly assigned to be in KIPP schools versus control schools. And it could be done: KIPP schools sometimes have waiting lists and use lotteries to pick students. Unchosen children could be assigned to a randomized control group and given the same tests as the KIPP students. KIPP itself is worried about picking the cream of the crop. To their credit, some schools have taken steps to ensure that they get as many underprivileged minority students as possible, the population they regard as their target.

But there is no question that there is a severe self-selection problem, because parents, not researchers, decide whether their kids will get into a KIPP school. This automatically means that some of the performance advantages of the KIPP students might come not from the school, but from having parents, and perhaps other advantages, that work to their benefit. The question is, could self-selection have produced the results that were obtained from the students?

As a start to answering this question, we can at least be reasonably sure that attrition is not a major contributor to the favorable results. Only about 9 percent of students at schools studied by SRI International exit each year, some to avoid being held back. The schools themselves discourage some students from continuing, though no one is ever actually expelled. But relatively few leave, so self-selection out of the school seems not to be a serious contributor to the high scores of KIPP children.

In short, though I do not doubt that students whose parents choose KIPP schools are more promising than their demographics would indicate, I cannot imagine that the results obtained by KIPP could be primarily due to self-selection. However, before society invests a huge amount of money in KIPP schools, some fully randomized studies should be conducted.

I have little doubt that such randomized studies will show that poor minority children—at any rate, those whose parents care enough to get them into KIPP schools—can perform academically at levels as high as those that approximate those of middle-class whites. The next step will be to find out whether children whose parents are not so concerned about their children's education can also benefit from KIPP-type programs.

High School Math for Poor Hispanics

You may have seen *Stand and Deliver,* the movie about math teacher Jaime Escalante's achievement in getting East Los Angeles barrio students—who typically did not graduate from high school—to pass AP calculus at higher rates than students at rich Beverly Hills High, and for that matter at most elite high schools in the country. But is the story told by the movie true?

There's good news and bad news about Escalante's feat. Most importantly, that it happened is perfectly true.

But unfortunately it did not happen in the way the movie implies: In real life Escalante did not announce to unsuspecting seniors that he was going to make them into math whizzes that year. He built up math programs at junior high feeder schools that brought highly prepared students into his three-year high school. And he made sure his students had excellent courses in high school math before they ever came into his class. But all that was accomplished with substantial opposition from the first principal he worked under. Things began to move smoothly only when a sympathetic principal came on board (and banned students from being on athletic teams if they maintained less than a C average). And then Escalante had to fight a teachers' union when his classes grew to be much larger than union rules allowed. He couldn't find enough good teachers to increase the number of classes taught, so classes grew too large.

Then the sympathetic principal was replaced by one less sympathetic. Escalante left his school over the class-size issue and other problems. The program became progressively less successful after his departure, though the high school continues to do far better with AP math than most schools of its type.

The importance of Escalante's achievement is very great. It serves as an existence proof for the contention that disadvantaged minority kids can function in math at a level far higher than the national average.

Inexpensive Interventions by Social Psychologists

Some of my fellow social psychologists have recently come on the education scene, bringing interventions that are very simple to carry out and extraordinarily cost-effective.

Many Americans believe that abilities are essentially fixed at birth: either you have math ability or you don't. Others believe abilities are highly susceptible to manipulation: if you work hard, you will be better at a given skill than if you don't. Carol Dweck and her coworkers have measured attitudes about ability in a group of mostly minority junior high school students, asking for beliefs about such questions as "You have a certain amount of intelligence, and you really can't do much to change it" and "You can always greatly change how intelligent you are." They showed, not surprisingly, that students who believe that ability is a matter of hard work get higher grades than students who believe ability comes from the genes.

Dweck and her colleagues then tried to convince a group of poor minority junior high school students that intelligence is highly malleable and can be developed by hard work. The thrust of the intervention was that learning changes the brain by form-ing new neurological connections and that students are in charge of this change process. Dweck reported that some of her tough junior high school boys were reduced to tears by the news that their intelligence was substantially under their control. Students exposed to the intervention worked harder, according to their teachers, and got higher grades than students in a control condi-tion. The intervention was more effective for children who ini-tially believed that intelligence was a matter of genes than it was for children who already were inclined to believe that it was a matter of hard work.

Joshua Aronson and his colleagues have performed similar experiments, with dramatic results. One study was conducted with poor minority students in Texas who were just beginning junior high school. Their intervention was intensive and the results were dramatic.

Each student in the Texas study was assigned a college-student mentor for their first year in junior high. The mentors discussed a variety of issues related to school adjustment. The mentors for the control participants gave information about drugs and encouraged their students to avoid taking them. Experimental-group mentors told their students about the expandable nature of intelligence and taught them how the brain can make new connections throughout life. Every student was exposed to a Web page that reinforced the mentor's message. For students in the experimental group, this Web site showed animated pictures of the brain, including images of neurons and dendrites, and provided narratives explaining how the brain forms new connections when new problems are being solved. The mentors also helped the students design a Web page in which the students presented, through words and pictures of their own making, the message that the mentor had been presenting.

The effects of the intervention were very powerful. On the math portion of the Texas Assessment of Academic Skills (TAAS), performance by male students exposed to the intervention was .64 SD higher than for males not exposed to the intervention. For females, who tend to have worries about whether their gender makes them less talented in math, the difference was 1.13 SDs. For reading, students exposed to the intervention did .52 SD better than students in the control group.

Daphna Oyserman and her coworkers set up an elaborate intervention with poor minority junior high school students. They ran several sessions designed to make the students think about what kind of future they wanted to have, what difficulties they would likely have along the way, how they could deal with those difficulties, and which of their friends would be most helpful in dealing with them. These were supplemented with sessions during which students worked in small groups on how to deal with everyday problems, social difficulties, academic issues, and the process of working toward high school graduation. The intervention had a modest effect on grade point average of .23 SD, a moderately large effect of .36 SD on standardized tests, and a very big effect on likelihood of retention in grade of .60 SD.

Small interventions can also make a difference in college. Most students worry about social acceptance and fitting in on campus, but for minority students these concerns can be particularly worrisome. If they fail to make friends, because there are not that many minority students on campus and because they may feel ill at ease with majority students, they may begin to wonder if they belong on campus. It is common for minority students' motivation to decline and for their grade point average to suffer as they go through school.

Social psychologists Gregory Walton and Geoffrey Cohen reasoned that lagging performance could be nipped in the bud if minority students knew that worries about social acceptance were common for all students, regardless of ethnicity, and that their situation would likely improve in the future. The researchers performed a modest intervention with black students at a prestigious private university. They invited black and white freshmen to participate in a psychology study at the end of their freshman year. The intent of the experimenters was to convince an intervention group that worries about social acceptance were common but tended to vanish as they developed more friendships. The experimenters expected that this would help black students to realize that the best way to understand their social difficulties was not in terms of their race ("I guess my kind of people don't really belong at this kind of place") and to replace that with the belief that their experiences were shared ("I guess everybody has these kinds of problems"). The researchers believed that recognition of their common problem—and its probable solution—would likely keep the students from worrying about belonging and help them focus on academic achievement.

To drive the point home, Walton and Cohen had students in the intervention group write an essay about the likelihood of improvement in their social situation in the future and deliver a speech in front of a video camera, which they were told would be shown to new students at the school "so that they know what college will be like." The investigators then measured academic achievement behavior over the next week, as well as the students' grade point average the subsequent semester.

The intervention had a big positive effect on blacks but not on whites. In the period after the intervention, blacks reported studying more, contacting professors more, and attending more review sessions and study group meetings. The subsequent term, grades of the blacks in the intervention group reflected these behaviors: their grades were a full standard deviation higher than those of blacks in the control group.

College as a Gap Reducer

It turns out that college itself has a huge effect on the intellectual abilities of blacks, substantially more than its effect on whites.

The black/white IQ gap grows in high school. Some hereditarians interpret this fact to indicate that the genes assert themselves more and more over the course of development. At each higher level of education, therefore, blacks could be expected to be farther and farther behind. Herrnstein and Murray, for example, maintained that it was unlikely that education beyond high school would serve to reduce the racial gap in IQ.

Data showing that the black/white gap grows during high school come from the National Longitudinal Survey of Youth, which administered the Armed Forces Qualification Test (AFQT). That test is a portion of the entire ability battery that the armed forces gives to potential soldiers. The AFQT correlates so highly with intelligence tests that it is appropriate to regard it as a measure of IQ. The test was given to participants in the survey at different points in their educational career from the age of fourteen to the age of twenty-one.

There is no question that, as Herrnstein and Murray showed, black ability increases less in high school than does white ability. The increase in the size of the gap is great enough to be quite disturbing. Blacks start out high school with scores less than three-fifths of a standard deviation lower on the AFQT than white scores, but end high school with scores almost a whole standard deviation lower than white scores.

Psychologist Joel Myerson and his coworkers set out to determine whether the same dismal failure of growth occurs for blacks in college. Based on the theory that ability differences manifest themselves more and more over time, we would expect less value added to IQ for blacks than for whites in college, resulting in a gap larger than that found in high school.

What Myerson and his colleagues found is the opposite. At the end of high school, black students who ultimately graduated from college performed more than 1 SD worse than white students who ultimately finished college. But the white students gained very little in IQ over the course of college, whereas the black students gained IQ at a remarkable clip, ending up with an average IQ only a little more than .40 SD below the average for whites. This difference in value added by a college education is huge.

Why do blacks gain so much in college? It may make more sense to ask why they gained so little in high school. The most obvious answer to that question is that blacks go to worse high schools than whites. A second answer is that pressure not to act white is harder to resist in high school than in college (if the pressure even exists at college).

A third possible answer lies in the research on stereotype threat, which demonstrates the remarkable variation in both test performance and motivation among black students, depending on the nature of the social circumstances they face. For example, Steele and Aronson's experimental research documented remarkably better test performance among black students when the test was presented in a nonthreatening way—that is, when test-takers weren't confronted with the explicit scrutiny of their intellectual abilities, presumably because doing so makes them anxious that their performance will mark them as fitting the stereotype of intellectual inferiority. In addition to impairment on tests, black high school students are more likely than white students to adapt to this evaluative discomfort in unhelpful ways, such as avoiding challenge and disengaging from academic pursuits, which are characteristic responses to stereotype threat seen among middle- and high-school-age students. One study that followed black students over their high school careers found a particularly sharp decline in black males' engagement with academics as they progressed through high school, so that by the time they were in twelfth grade there was no connection between their feelings of self-worth and their academic achievement. These responses seem particularly likely among students who buy into the stereotypes about their group. Another long-term study found a direct link between students' worries that the negative stereotypes about their group might be true and their later reduction in effort. So the circumstances that maximize stereotype threat may be especially prevalent in high school.

But the truth is, we do not know the reasons for the black gains in college. What we do know is that college produces a huge reduction in the ability gap. We also know that we have yet another important piece of evidence contradicting the idea that the IQ gap increases with age because of a genetic deficiency that manifests itself more over time. The IQ gap actually narrows very substantially over the college years.

Summing Up

So what do we know about intervention with minority children and the poor? Several very striking things. Perhaps the main lesson is that what works and what doesn't is an empirical question.

Some early childhood programs that seem very reasonable do not have very big—or very lasting—effects. Head Start is a very sensible program that ought to make a big difference—and it does at first. But when students are left in poor families, in poor neighborhoods, and in poor schools, the Head Start gains in IQ fade and the academic achievement effects are not very large. But there are pre-kindergarten programs that have huge initial effects on IQ, along with significant residual effects on IQ if its graduates attend high-quality public schools. Even when children are not in good schools, the best programs nevertheless have very large academic achievement effects and enormous social benefits in terms of reduced crime and welfare dependence. Pre-kindergarten programs in general benefit poor kids and minority kids more than better-off white kids.

A similar pattern exists for elementary and junior high school interventions. I've described many programs that do not have much effect on student achievement and therefore are unlikely to be able to close the racial and SES gaps. The factors that do make a difference are the quality of the individual teacher, teacher incentives (probably, though much more work needs to be done to find out just what kinds of incentive programs work best and are most practical), and class size (which seems to have a bigger effect on black kids than on white kids).

Some teaching techniques discussed in Chapter 4 that are not terribly expensive are known to have fairly sizable effects and might help to reduce the achievement gap. These include computer programs to teach math and writing (which might actually be less expensive than standard instructional methods) and "cooperative learning" situations (which would add little if anything to cost). Even if these techniques were not found to reduce the gap, the fact that they improve the achievement of children in general indicates that they should be used for educating all kinds of students. A rising tide is welcome even if it lifts all boats equally.

We know that one math program—SEEDS—and one reading program—Reading Recovery—have big effects on minority kids. Reading Recovery is particularly inexpensive and provides a big bang for the buck.

A host of whole-school interventions have been tried, and the results are mostly disappointing. However, there is one hugely important exception—the KIPP program. Like Jaime Escalante, KIPP teachers can get poor minority pupils to perform at or above the level typical of white middle-class pupils. And that's true even when they start the program late: surprisingly, the effects are big for students who start it in fifth grade. It remains to be seen what KIPP could accomplish if it were to start with much younger kids. Let's hope we find out soon.

Paying for Gap Reduction

But can we afford the effective programs? A better question to ask is, can we afford not to have them? Many economists have evaluated the benefit-cost ratio of the most successful pre-kindergarten interventions. The Nobel Prize–winning economist James Heckman estimates the payback of the Perry Preschool Program—in terms of special education classes avoided, extra years of schooling avoided, crime and welfare costs avoided, and higher incomes for its participants—to be eight to one. This is equivalent to a return on investment of 17 percent per year. And that is just the cold monetary calculation. The gains to participants' quality of life and those of their families and neighbors are not even included in those calculations. The initial cost of the Perry program is high—estimates range between $12,000 and $16,000 per student in 2007 dollars. It is much more expensive than Head Start (though the costs of future Perry-type programs should be lower because there would be no research component, which was a significant part of the Perry program costs) but it would be vastly more effective. The same is true of the Abecedarian intervention, which has been calculated as having a benefit-cost ratio of $3.78 to the dollar. Even when benefits are calculated strictly to the taxpayer, in terms of education, welfare, and criminal justice charges saved, the costs of the most successful pre-kindergarten programs are repaid in time.

So how much would it cost to enroll, say, the poorest third of children in a Perry or Abecedarian program from birth to the time of entry into kindergarten? There are about 7 million such children in the United States, and either the Perry or the Abecedarian program might cost as much as $15,000 per child per year. That works out to about $105 billion. But to soften the sticker shock, let me quickly point out that about $20 billion in public money is currently spent on pre-kindergarten programs. An unknown amount of privately spent money should also be subtracted. So should the additional earnings by the mother in the years during and after the program. The ultimate benefits to the child and to society should also be subtracted. Moreover, not nearly enough studies have been done on the most expensive programs—it is possible that very substantial improvements in cognitive and social functioning could be achieved by spending notably less than $15,000 per child. Finally, recall that gains in pre-kindergarten programs are proportional to need. The lower the mother's IQ and SES, the greater the gains. So even if we provide intensive pre-kindergarten to only the neediest one-sixth or one-twelfth of the under-five population, the payoff would be very great.

As a yardstick for measuring the gross initial cost of early childhood education, note that the projected post-2001 tax cuts for the richest 1 percent of U.S. families will cost the treasury $94 billion in 2009 alone.

How about the apparently extremely effective KIPP programs? KIPP schools are actually not much more expensive than the regular public schools. (Some KIPP schools cost less than the public schools in their districts.) But KIPP gets its results by dint of the willing slave labor of its young, idealistic teachers who are paid only slightly more than public school teachers of comparable experience. And the KIPP teachers cannot keep up the pace for many years. Not surprisingly, unions have begun to oppose KIPP schools for their rate-busting.

How much would it cost to run KIPP schools if teachers were paid the same rate per hour that public school teachers earn— which seems not only fair but necessary if sufficient teachers are to be found? KIPP students get about 60 percent more contact time with their teachers. Most of the cost of education is in the form of physical plant, administrative and maintenance costs, and interest on debt, which are not increased by KIPP techniques. The cost per pupil in the United States for the average public school was about $8,000 in 2005, and about a third of this is for teachers' pay. If we assume that teachers' pay would be 60 percent more, and if we assume that a KIPP-type program would be made available to one-third of the 40 million children five to fourteen years old, it would cost an extra $35 billion. But again, these costs would be offset to a significant degree by child-care costs saved plus additional earnings by the mother. The economic gains over the lifetime for children in such schools cannot be calculated at this point. At a minimum they could be expected to defray a significant part of their extra cost.

To be clear, I am not advocating instituting particular programs at this time. A huge amount of research needs to be done to establish whether something like the Perry or Milwaukee or Abecedarian program would be effective and feasible if scaled up to national proportions, and the same thing is true of KIPP-type programs. In the case of KIPP, we would have to see how much gain could be expected for children of parents who did not exert effort to get their children into the program and keep them there.

We have existence proofs, however, that a marked reduction in the IQ and achievement gaps is possible, and we know that the costs of the effective interventions are at least conceivable. It would be irresponsible to fail to do the necessary research to find out what kinds of intensive programs do the most good.

Finally, if we want to make the poor smarter, a good way to do it might be to make them richer. The Scandinavian countries are much more egalitarian in their income distribution than the United States is, and the achievement gap between their richest and poorest children reflects that relative equality. Honest employment in a job having social value should pay enough to support a family. This could be achieved in part by increasing the minimum wage (which even with the new increases will be only 73 percent of what it was forty years ago), the Earned Income Tax Credit, and child tax credits.

The economic cost of at least some of this—and probably even more than the cost—would be recouped by increasing the productivity of the poor and reducing crime and welfare rates. We would likely do well by doing good.

CHAPTER EIGHT

Advantage Asia?

Good grief, those scores are positively Asian.
—One European-American Silicon Valley high school senior to
another, upon hearing about her extremely high SAT scores

*If there is no dark and dogged will, there will be no shining
accomplishment; if there is no dull and determined effort,
there will be no brilliant achievement.*   —Chinese saying

HERE ARE SOME STATISTICS that should serve to concentrate the minds
of people of European descent.

- In 1966, Chinese Americans who were seniors in high school
  were 67 percent more likely to take the SAT than were European
  Americans. Despite being much less highly selected, the Chinese
  Americans scored very close to European Americans on average.
- In 1980—when they were thirty-two years old—the same Chinese Americans from the "class of '66" were 62 percent more
  likely to be in professional, managerial, or technical fields than
  were European Americans.
- In the late 1980s, the children of Indochinese boat people
  constituted 20 percent of the population of Garden Grove in
  Orange County, California, but claimed twelve of fourteen high
  school valedictorians.
- In 1999, U.S. eighth-graders scored between .75 and 1.0 SD
  below Japan, Korea, China, Taiwan, Singapore, and Hong Kong
  in math and between .33 and .50 SD below those countries in
  science as indicated by the Third International Mathematics and
  Science Study.

- Although Asian Americans constitute only 2 percent of the
  population, all five of the Westinghouse Science Fair winners in
  2008 were Asian Americans.
- Asian and Asian American students now constitute 20 percent
  of students at Harvard and 45 percent at Berkeley.

So European Americans might as well throw in the towel.
Asians are just plain smarter.

Actually, probably not. At least not as indicated by traditional
IQ tests. Herrnstein and Murray, Rushton and Jensen, Philip Vernon, Richard Lynn, and others have reported that there are IQ
differences favoring Asians, but Flynn has shown that such reports
are due in good part to the failure of the researchers to report
Asian IQs based on contemporary IQ test norms rather than on
outmoded norms and to using small and unrepresentative samples. Basing scores on outmoded norms has the effect of errone-
ously raising Asian IQs. Flynn reviewed sixteen different studies,
the results of which were fairly consistent with one another. Most
showed that East Asians had slightly lower IQs than Americans.

What is not in dispute is that Asian Americans achieve at a level
far in excess of what their measured IQ suggests they would be
likely to attain. Asian intellectual accomplishment is due more to
sweat than to exceptional gray matter.

The Asian Drive for Achievement

Harold Stevenson and his coworkers studied the intellectual abilities and school achievement of children in three different cities
chosen to be highly similar socioeconomically: Sendai in Japan,
Taipei in Taiwan, and Minneapolis in the United States. They
measured the intelligence and reading and math achievement of
random samples of children in the first and fifth grades. We cannot know if the IQ tests really provided measures of intelligence
that are fully comparable across the three populations (though the
researchers believed they did—and make a pretty good case for

that). Nevertheless, in the first grade, the Americans outperformed the Japanese and the Chinese on most intelligence tests. The authors attributed this to the greater effort of American parents to stimulate their preschool kids intellectually. Whatever the reason for the high American performance in the first grade, by the fifth grade the superiority of American children in IQ was gone. From these sets of facts we learn that regardless of who was smarter than whom in the first grade, the Americans had lost considerable ground to the Asian children by the fifth grade.

But the truly remarkable finding of this study was that math achievement of the Asian students was leagues beyond that of the U.S. students. The identical problems were given to Japanese, Taiwanese, and American children. By the fifth grade, Taiwanese children scored almost 1 SD better in mathematics than American children, and the Japanese scored 1.30 SDs better than American children. Even more astonishing, in a more extended study, Stevenson and his coworkers looked at the math performance of fifth-graders in many different schools in China, Taiwan, Japan, and the United States. There wasn't a lot of difference among the Asian countries. Schools in all three countries performed at about the same level. There was more variability among the U.S. schools. But the very best performance by an American school was equal to the worst performance of any of the Asian schools!

IQ is not the point: something about Asian schools or the motivation of Asian children differs greatly from American schools or American children's motivation.

Let's start with the schools. Children in Japan go to school about 240 days a year, whereas children in the United States go to school about 180 days a year. The Asian schools are probably better, but the performance of Asian American children in U.S. schools shows that Asian motivation counts for an awful lot.

The Coleman report on educational equality in the United States, published in 1966, measured the intelligence of a very large random sample of American children, and Flynn followed them until they were thirty-six years old on average. Americans of East

Asian descent scored about 100 on the nonverbal portion of IQ tests and about 97 on the verbal portion, so they had a slightly lower overall IQ than did Americans of European descent.

Despite their slightly inferior performance on IQ tests, the Chinese Americans of the class of 1966 were about half as likely as other children to have to repeat a grade in K–12. Foreshadowing things to come, when the Chinese American children were compared with European American children in grade school, they did slightly better on achievement tests. By the time they were in high school, the Chinese Americans were scoring one-third of a standard deviation higher than European Americans on achievement tests. At a given IQ level, the Chinese Americans performed one-half of a standard deviation higher on typical achievement tests, compared with European Americans. The overachievement was particularly great on mathematics tests. In tests of calculus and analytic geometry, the Chinese Americans surpassed European Americans by a full standard deviation. When students were seniors in high school, the Chinese Americans performed about one-third of a standard deviation better on SAT tests than did Americans of the same IQ.

By the age of thirty-two, the determination of the Chinese Americans in the class of 1966 had paid a double dividend. To get the educational credentials to qualify for professional or technical or managerial occupations, they needed a minimum IQ of 93, compared to 100 for whites. More important, of those with the IQ to qualify, fully 78 percent had the persistence to get their credentials and enter those occupations compared to 60 percent of whites. The resulting total dividend was 55 percent of Chinese Americans in high-status occupations, compared to a third of whites. The number for Japanese Americans was about halfway between the numbers for these two groups.

Flynn found similar overachievement relative to IQ on achievement tests and in occupations in a wide variety of studies of East Asians.

Notice that the overachievement of Asian Americans, as indicated by the marked difference between measured IQ and academic achievement, is sufficient by itself to establish that achievement tests such as those given in K–12 classrooms and the SAT are not merely IQ tests by another name. They measure intellectual achievement as opposed to the power of memory, perception, and reasoning of the kind that IQ tests measure. Note also that the overachievement of Asian Americans establishes that academic achievement can be a better predictor of ultimate socio-economic success than IQ.

Recently, Flynn studied the children of the members of that original class of 1966. Since we know that being raised in homes of higher social class is associated with higher IQ, we would expect the children to have higher IQs than not only their own parents but also the population at large. And indeed they did. The mean of the Chinese American children when they were preschoolers was 9 points higher than the white average. But then most went to ordinary American schools, which we would expect would not be ideal for their intellectual development. In fact, the average of their IQs steadily declined until it was only 3 points above the white mean by the time they were adults.

Notice the arbitrariness of describing what Asians accomplish as overachievement. I used the phrase "Asian overachievement" to a Korean friend who had just spent a year in the United States, where his children attended public schools. "What do you mean by 'Asian overachievement'?," he expostulated. "You should say 'American underachievement'!" He told me that he was astonished when he attended ceremonies at the end of the year for his daughter's school and discovered that an award was given for having done all of the homework assignments. His daughter was one of two recipients of the award. To him, giving an award for doing homework was about as preposterous as giving an award for eating lunch. It is taken absolutely for granted by Asians. He is right to insist that the phenomenon is one of American under-achievement. It's quite reasonable to regard high achievement as the default state of affairs and what most Americans do as slacking to one degree or another.

My Korean friend's bemusement touches on the key to understanding Asian achievement magic.

Asian and Asian American achievement is not mysterious. It happens by working harder. Japanese high school students of the 1980s
studied 3½ hours a day, and that number is likely to be, if anything,
higher today. The high-school-age children of the Indochinese boat
people studied 3 hours a day. American high school students in
general study an average of 1½ hours a day. (Black eighth-grade
children in Detroit study, on average, 2 hours per *week*. Of course,
at least some of this failure to do homework would have to be
attributed to a school milieu that does not expect much.)

There is also no mystery about why Asian and Asian American
children work harder. Asians do not need to read this book to find
out that intelligence and intellectual accomplishment are highly
malleable. Confucius set this matter straight twenty-five hundred
years ago. He distinguished between two sources of ability, one by
nature—a gift from Heaven—and one by dint of hard work.

Asians today still believe that intellectual accomplishment—
at any rate, doing well in math in school—is primarily a matter
of hard work, whereas European Americans are more likely to
believe it is mostly a matter of innate ability or having a good
teacher. Asian Americans have attitudes on this topic that are in
between those of East Asians and European Americans.

Asians and Asian Americans have another motivational advantage over Westerners and European Americans. When they do
badly at something, they respond by working harder at it. A team
of Canadian psychologists brought Japanese and Canadian college
students to a laboratory and had them work on creativity tests.
After the study participants had been working on them for a while,
the researchers thanked them and told them about how well they
did. Regardless of how well they had actually done, the researchers told some of the participants that they had done very well and
others that they had done rather badly. The investigators then gave
the participants a similar creativity test and told them to spend as
much time as they wanted on it. The Canadians worked longer on

the creativity test if they had succeeded on the first one than if they had done badly, but the Japanese worked longer on the creativity test if they had failed on the first one than if they had succeeded.

Persistence in the face of failure is very much part of the Asian tradition of self-improvement. And Asians are accustomed to criticism in the service of self-improvement in situations where Westerners avoid it or resent it. For example, Japanese schoolteachers are observed in their classrooms for at least ten years after they begin teaching. Their fellow teachers give them feedback about their teaching techniques. It is understood in Japan that you cannot be a good teacher without many years of experience. In the United States, we tend to toss teachers into the classroom and assume they can do a good job from the get-go. Or if not, it's because they haven't got what it takes.

But a still more important reason for Asians making the most of their natural intelligence is that their culture—as channeled to them by their families—demands it. In the case of Chinese culture, the emphasis on academic achievement has been present for more than two thousand years. A bright Chinese boy who worked hard and did well on the mandarin exams could expect to elevate himself to a well-paying high government position. This brought honor and wealth to his family and his entire village—and the hopes and expectations of his family and fellow villagers were what made him do the work. There was substantial upward mobility via education in China a couple of millennia before this was the case in the West.

So Asian families are more successful in getting their children to achieve academically because Asian families are more powerful agents of influence than are American families—and because what they choose to emphasize is academic achievement.

Eastern Interdependence and Western Independence

Why should Asian families be such powerful agents of influence? Here I need to step back a bit and note some very great differences

between Asian and Western societies. Asians are much more interdependent and collectivist than Westerners, who are much more
independent and individualist. These East-West differences go
back at least twenty-five hundred years to the time of Confucius
and the ancient Greeks.

Confucius emphasized strict observance of proper role relations
as the foundation of society, the relations being primarily those of
emperor to subject, husband to wife, parent to child, elder brother
to younger brother, and friend to friend. Chinese society, which
was the prototype of all East Asian societies, was an agrarian
one. In these societies, especially those that depend on irrigation,
farmers need to get along with one another because cooperation is
essential to economic activity. Such societies also tend to be very
hierarchical, with a tradition of power flowing from the top to the
bottom. Social bonds and constraints are strong. The linchpin of
Chinese society in particular is the extended family unit. Obedience to the will of the elders was, and to a substantial degree still
is, an important bond linking people to one another.

This traditional role of the family is still a powerful factor in the
relations of second- and even third-generation Asian Americans
and their parents. I have had Asian American students tell me that
they would like to go into psychology or philosophy but that it
is not possible because their parents want them to be a doctor or
an engineer. For my European American students, their parents'
preferences for their occupations are about as relevant to them as
their parents' taste in art.

The Greek tradition gave rise to a fundamentally new type of
social relations. The economy of Greece was based not on large-
scale agriculture but on trade, hunting, fishing, herding, piracy,
and small agribusiness enterprises such as viniculture and olive
oil production. None of these activities required close, formalized relations among people. The Greeks, as a consequence, were
independent and had the luxury of being able to act without being
bound so much by social constraints. They had a lot of freedom
to express their talents and satisfy their wants. The individual

personality was exalted and considered a proper object of commentary and study. Roman society continued the independent, individualistic tradition of the Greeks, and after a long lull in which the European peasant was probably little more individualist than his Chinese counterpart, the Renaissance and then the Industrial Revolution took up again the individualist strain of Western culture and even accelerated it.

It is hard for someone steeped only in European culture to comprehend the extent to which achievement in the East is a family affair and not primarily a matter of individual pride and status. Like the ancient candidate for mandarin status, one achieves because it is to the benefit of the family—both economically and socially. Although there may be pride in personal accomplishment, achievement is not primarily a matter of enriching oneself or bringing honor to oneself.

And—here's the big advantage of Asian culture—achievement for the family seems to be a greater goad to success than achievement for the self. If I, as an individual Western free agent, choose to achieve in order to bring myself honor or money, that is my decision. And if I decide that my talents are too meager or I don't want to work hard, I can choose to opt out of the rat race. But if I am linked by strong bonds to my family, and fed its achievement demands along with my meals, I simply have no choice but to do my best in school and in professional life thereafter. And the demand is reasonable because it has been made clear to me that my achievement is a matter of will and not just innate talent.

The achievement advantage of Asian Americans over European Americans is likely to increase. Prior to 1968, Asian immigrants to the United States were probably not more natively intelligent than their compatriots who remained at home. But the immigration laws of the 1960s make it relatively easy to come to the United States for a person who is a professional and relatively hard for someone who is not. The Asian American newcomers are going to have a cultural advantage over European Americans in general because they are professional and managerial types as well as being

East Asians. Both their social class and their culture are going to be favorable for the maximum development of educational and professional success of their children. And their children are going to have a genetic advantage as well because of the selection for talent. (This genetic advantage is likely to be slight. As we will see in the next chapter, environmental bottlenecks do not have much effect on the IQ of generations after the bottleneck.)

Holistic and Analytic Habits of Thought

The cultural differences of East and West result not just in quantitative differences in intellectual achievement but also in qualitative differences in habits of mind. Effective functioning for East Asians depends on integrating one's own desires and actions with those of others. *Harmony* has been the watchword for social relations for twenty-five hundred years in China. Effective functioning for Westerners is not so dependent on dealing with others. Westerners have the luxury of acting independently of the wishes of other people.

These social differences have given rise to habits of mind on the part of Easterners that I describe as *holistic*. Easterners pay attention to a wide range of objects and events; they are concerned with relationships and similarities among those objects and events; and they reason using dialectical forms of thought, which includes finding the "middle way" between opposing ideas. Western perception and thought are *analytic*, which is to say that Westerners focus on a relatively small part of the environment, some object or person that they wish to influence in some way; they attend to the attributes of that small part with a view toward categorizing it and modeling its behavior; and they often reason using formal rules of logic.

The need to attend to others means that the perception of Easterners is directed outward to a broad swath of the social environment and, as a consequence, to the physical environment as well. Takahiko Masuda and I showed people brief animated films of

underwater scenes and then asked them to tell us what they had
seen. Take a look at Figure 8.1, which shows a still photo taken
from one of the films. The Americans focused primarily on the
most salient objects—large, rapidly moving fish, for example. A
usual first response would be, "I saw three big fish swimming off
to the left; they had pink spots on their white bellies."

*Figure 8.1. Still photo from a color animation film shown to
Japanese and Americans who were asked to report what they saw.
From Masuda and Nisbett (2001).*

The Japanese reported seeing much more of the environment—
rocks, weeds, inanimate creatures such as snails. A typical initial
response would be, "I saw what looked like a stream; the water
was green; there were rocks and shells on the bottom." In addition
to paying attention to context, the Japanese noticed relationships
between the context and particular objects in it. For example,
they were inclined to note that one object was next to another or
that a frog was climbing on a plant. Altogether, the Japanese were
able to report 60 percent more details about the environment than
were Americans.

In another study, Masuda demonstrated that when shown cartoon pictures of a central figure flanked by other people and asked to judge the mood of the central figure, the Japanese were much more influenced in their judgments by the expressions on the surrounding faces than were the Americans.

Asians and Westerners see different things because they are looking at different things. My coworkers and I have rigged people up with devices that can measure what part of a picture they are looking at every millisecond. Chinese spend more time looking at the background than do Americans and make many more eye movements back and forth between the most salient object and the background.

The greater attention to context allows East Asians to make correct judgments about causality under circumstances where Americans make mistakes. Social psychologists have uncovered what they call "the fundamental attribution error." People are prone to overlook important social and situational causes of behavior and attribute the behavior instead to what they assume are attributes of the actor—personality traits, abilities, or attitudes. For example, when reading an essay that an instructor in a course or an experimenter in a psychology study has asked someone to write in favor of capital punishment, Americans assume that the writer must hold the view that he expressed. And they do this even when the experimenter has just requested that they write an essay upholding a view the experimenter chose. Koreans in this situation correctly make no assumption that the person whose essay they read actually holds the position he takes in the essay.

The greater attentiveness to context has been characteristic of East Asians since the time of the ancient Chinese, who understood the concept of action at a distance. This made it possible for them to understand the principles of magnetism and acoustics and allowed them to figure out the true cause for the tides (which escaped even Galileo). Aristotle's physics, in contrast, was completely focused on the properties of objects. In his system, a stone fell to the bottom when dropped in water because it had the property of gravity, and a stick of wood floated on the water because it had the property of levity. There is no such property as levity, of course, and gravity is not found in objects but in the relation between objects.

Despite the greater correctness of ancient Chinese physics, and despite the fact that China was leagues ahead of the Greeks in technological achievements, it was the Greeks who invented formal science. Two things made this possible for the Greeks.

First, because the Greeks were fixated on objects, they were concerned with the attributes of objects and with determining the categories to which they belonged. In order to understand the behavior of objects, the Greeks invented rules that presumably governed the behavior of objects. And rules and categories are what constitute science at base. Without them, there can be no explicit, generalizable models of the world to test. There can only be technology, no matter how sophisticated.

Second, the Greeks invented formal logic. As the story goes, Aristotle had gotten impatient with hearing poor arguments in the marketplace and the political assembly and so came up with logic in order to rule out forms of argument that are defective. In any case, logic does in fact serve that function in the West.

Logic was never of much interest in China. In fact, it appeared just once, briefly, in the third century BC, and it was never formalized. The Greeks could invent logic precisely because their habit of argumentation was socially acceptable. In ancient China, and in most of East Asia today, disagreements are a risky business—you might make an enemy if you contradict another person's point of view. Instead of logic, the abstract reasoning patterns of the East tend toward dialecticism, including a concern with finding the "middle way" between opposing arguments and an emphasis on integrating different points of view.

Like rules, categories, and explicit models, formal logic is an extremely helpful tool for science. But the Greeks went overboard in their fondness for logical argument. They rejected the concept of zero because, they reasoned, zero was equivalent to "non-being" and non-being cannot be! And Zeno's famous paradoxes are the result of logic gone wild. (For example, motion is impossible. For an arrow to reach a target, it would have to go half the distance between the bow and the target, then half that distance, and so on ad infinitum, and thus could never reach its target. This strikes us as comical, but the Greeks thought this was a real stumper.)

Social practices and habits of thought tend to get ingrained, and so contemporary social and cognitive differences between East and West are much like those of ancient times. Thus we might expect Westerners to be more likely to emphasize rules, categories, and logic, and Easterners to be more likely to emphasize relationships and dialectical reasoning. And, in fact, my coworkers and I find this to be the case.

When we presented people with three words such as *cow*, *chicken*, and *grass*, and asked them which two go together, we got very different answers from Easterners and Westerners. Americans were more likely to say *cow* and *chicken* go together because they are both animals; that is, they belong to the same taxonomic category. Asians, however, focusing on relationships, were more likely to say that *cow* goes with *grass* because a cow eats grass.

We also presented syllogisms to Americans and Asians and asked them to judge the validity of their conclusions. We found that Asians are just as good as Americans at judging the validity of syllogisms that are stated in abstract terms—all *A*s are *X*, some *B*s are *Y*, and so on—but are likely to be led astray when dealing with familiar content. Asians are inclined to judge conclusions that follow from their premises to be invalid if they are implausible (e.g., All mammals hibernate/rabbits do not hibernate/rabbits are not mammals). And Asians are likely to judge as valid conclusions that are in fact invalid but which are plausible.

Finally, it is possible to show that Americans sometimes make mistakes in reasoning owing to the same kind of "hyperlogical" stance that characterized the ancient Greeks. My coworkers and I showed that Americans will sometimes judge a given plausible proposition to be more likely to be true if it is contradicted by a less plausible proposition than if it is not contradicted. The Americans assume that if there is an apparent contradiction between two propositions, the more plausible one must be true and the less plausible one must be false. Asians make the opposite error of judging a relatively implausible proposition to be more likely to be true if it is contradicted by a more plausible proposition than if it is not contradicted—because they are motivated to find truth in both of two opposing propositions.

These perceptual and cognitive differences rest on brain activity that differs between Easterners and Westerners. For example, when Chinese are shown animated pictures of underwater scenes, an area of the brain known to respond to backgrounds and contexts is more active than it is for Americans. Conversely, an area of the brain known to respond to salient objects is less active for Asians than it is for Americans. Another brain-function study pursued the fact that Americans find it easier to make judgments about objects while ignoring their contexts, and East Asians find it easier to make judgments about objects which take into account their context. Consistent with this fact, regions of the frontal and parietal cortices that are known to be involved in attention control are more active when a person makes judgments of the nonpreferred, more difficult kind—that is, judgments taking into account contexts for Americans and judgments that require ignoring contexts for East Asians.

How do we know that these differences in perception and thought are social in origin and not genetic? There are two main reasons. First, in several of the studies we conducted, we compared Asians, Asian Americans, and European Americans. In all the studies the Asian Americans perceived and reasoned in ways that were intermediate between Asians and European Americans, and were usually more similar to those of European Americans. Second, Hong Kong is known to be a bicultural society, with Chinese customs mingling with English ones. We found residents of Hong Kong to reason in a fashion intermediate between how Chinese and European Americans reason. And when Hong Kong residents were asked to make causal attributions about the behavior of fish, they reasoned like Chinese after being shown pictures such as temples and dragons and like Westerners after being shown pictures such as Mickey Mouse and the U.S. Capitol!

Eastern Engineers and Western Scientists?

The different social inclinations and thought patterns of Easterners and Westerners have implications for doing well in engineering versus science.

Everyone has heard the cliché that Japanese make good engineers but lag in science. This is no mere stereotype. Japanese prowess in engineering is the wonder of American industry. And my colleagues who teach engineering and my friends who hire engineers tell me that not only are there more Asian American engineers per capita but they also make better engineers than European Americans on average.

However, in the decade of the 1990s, forty-four Nobel Prizes in science were awarded to people living in the United States, the great majority of whom were Americans, and only one was awarded to a Japanese. This is not entirely the result of a difference in funding. The Japanese have spent roughly 38 percent as much on basic research as have Americans over the last twenty-five years, and they spend twice as much as do Germans, who won five Nobel Prizes in the 1990s. China and Korea have been relatively poor, developing countries until recently, and it is too early to tell how successful their citizens will be in basic science. But it is possible to point to some of the roadblocks in the path to scientific productivity that might apply to all interdependent peoples who are inclined to be holistic thinkers.

First, several social differences between East and West favor Western progress in science. In Japan, which is more hierarchically organized than the West in many respects, and which places a greater value on respect for elders, more research money goes to older, no-longer-productive scientists. I believe that the premium on individual achievement and the respect for personal ambition in the West favors scientific accomplishment. Long hours in the lab do not necessarily do much for the scientist's family, but they are essential to personal fame and glory. Debate is taken for granted in the West and is regarded as an essential part of the scientific enterprise, but it is considered rude in much of the East. A Japanese scientist recently reported on his amazement at seeing American scientists who were friends sharply disagree with one another—and in public. "I worked at the Carnegie Institution in Washington, and I knew two eminent scientists who were good friends, but once it came to their work, they would have severe debates, even in the journals. That kind of thing happens in the United States, but in Japan, never."

Second, the Confucian tradition, of which Japan and Korea are a part, has little use for the idea that knowledge is valuable for its own sake. This starkly contrasts with the ancient Greek philosophical tradition, which prized such knowledge above all other kinds. (I emphasize the term *philosophical tradition* in the preceding sentence. There is an amusing passage in *The Republic* where an Athenian businessman castigates Socrates for his pursuit of abstract knowledge, telling him that although it is admittedly attractive in the young, it is disgusting in a grown man.)

Third, logic, the intellectual tool of debate, is more readily applied to real-world content by Westerners than by Easterners. Even the occasional hyperlogical habits of Westerners can be useful in science, however clumsy and even comical they can be in everyday life. Related to logic is the Western type of rhetoric found in formal discourse in science, law, and policy analysis. This consists of an overview of what is being discussed, general concerns about the topic, specific hypothesis, operations to test the hypothesis, discussion of pertinent facts, defense against possible counterarguments, and summary of conclusions. Training in this pattern of argumentation begins in nursery school: "This teddy bear is my lovey, I like him because . . ." Perhaps due to its roots in debate and formal logic, the Western form of rhetoric is not common in the East. I find with my own East Asian students that the standard rhetorical form is the last thing they learn on their way to a PhD.

Finally, there is the matter of curiosity. For whatever reason, Westerners seem to be more curious than Easterners. It is Westerners who have explored the Earth and immersed themselves in science and who regard the proper study of philosophy to be the fundamental nature of humankind. I do not know why this should be, though I can speculate on one possible source. We know that Westerners are constantly building causal models of the world. In fact, the children of Japanese who are living in the United States for business reasons are often regarded by American teachers as having weak powers of analysis because they do not build such causal models. One consequence of building explicit models is the element of surprise. The models lead to predictions that turn out to be wrong. This makes a person eager to get more accurate views—and more curious.

None of the habits of mind that are more characteristic of Easterners pose insurmountable obstacles to scientific excellence. The practice of science encourages mental patterns I have labeled as Western advantages, and the more steeped in scientific culture that Easterners become, the more natural will scientific habits of mind become. And Easterners may well be able to shape their distinctive habits of mind in ways that will provide advantages for scientific inquiry. Quantum theory in physics rests on contradictions that are anathema to the Western mind but congenial to the Eastern mind. Nils Bohr credited his deep knowledge of Eastern philosophy with his ability to generate quantum hypotheses.

For the time being, Westerners' advantage in science may be their ace in the hole in their friendly competition with the East. But don't count on the advantage lasting for long. Until fairly far into the last century, European scientists were puzzled by the failure of Americans to produce much science of note.

CHAPTER NINE

People of the Book

> *A man should sell all he possesses in order to marry the daughter of a scholar, as well as to marry his daughter to a scholar.* —The Talmud: Pesahim 49a
>
> *[The Jews] are peculiarly and conspicuously the world's intellectual aristocracy.* —Mark Twain, in a letter written in 1879
>
> *The United States today is the greatest fistic nation in the world, and a close examination of its 4,000 or more fighters shows that the cream of its talent is Jewish.*
> —Boxing announcer Joe Humphreys in 1930

IN AD 64, the Jewish high priest Yehoshua ben Gamla issued an edict specifying that all males be able to read the Talmud. This requirement was met within a hundred years after its promulgation. The next national group to achieve universal male literacy did not do so until about seventeen hundred years later.

Jewish Accomplishment

It scarcely seems coincidental that the people who first achieved literacy should today be extraordinarily distinguished intellectually. Ashkenazi Jews (those of European descent) are overrepresented among Nobel Prize winners by a factor ranging from 50 to 1 (the prize for peace) up to 200 to 1 (the prize for economics) in relation to the percentage of the world's population that is Jewish. It would be fairer, though, to compare the proportion of Jewish prize winners in the Western world, or even the United States. American Jews have received between 27 and 40 percent of all Nobel Prizes in science awarded to Americans (depending on whether we count as Jewish only individuals whose mothers and fathers are Jewish or all winners with at least one-half Jewish heritage). Jews represent less than 2 percent of the American population, so this constitutes an overrepresentation of about 15 to 1 (using the more conservative definition of who is Jewish). Approximately the same overrepresentation applies to American winners of the A. M. Turing Award for contributions to computing. And 26 to 34 percent (depending on how we count who is Jewish) of Fields Medals in mathematics for Americans have gone to Jews.

In the United States, Jews comprise 33 percent of Ivy League students, an approximately equal percentage of the faculty at elite colleges, and about 30 percent of Supreme Court law clerks. These are overrepresentations by a factor of 15 or more.

Jewish achievement is not limited to the purely intellectual. According to the 1931 census of Poland, Jews comprised 9.8 percent of the population. Jews, however, owned 22.4 percent of the wealth in the country. In the first four years after World War I, more than 70 percent of business licenses were issued to Jews. By 1929, Jews owned 45 percent of large and medium-sized commercial enterprises. By 1938, the proportion was 55 percent. By the mid to late 1930s, a majority of the owners of the following industries were Jews: textiles, chemicals, food, transportation, building materials, and paper.

So Jews have proved to be extraordinarily successful in many endeavors where intelligence is an advantage, including business and commerce.

Jewish IQ

The stereotype of Jews as highly intelligent, then, is certainly backed up by the statistics. It is also backed up by IQ tests. Jews have the highest average IQ of any ethnic group for which there are reliable data. Most reports place the average Ashkenazi Jewish IQ at two-thirds to one standard deviation above the white average. That is equal to an IQ of 110 to 115.

The degree of overrepresentation of Jews in intellectual realms is actually greater than would be expected on the basis of Jewish IQ. Let's arbitrarily specify that an IQ of 140 or above constitutes the average IQ at the most stratospheric levels of intellectual achievement—Nobel Prizes and the like. If we assume an average IQ of 110 for Jews, Jews would be expected to be overrepresented at IQ 140 by a factor of only 6 to 1. This is substantially less than the overrepresentation of at least 15 to 1 that is actually found, and very substantially less than that if we count people with one Jewish parent. Even if we assume an average Jewish IQ of 115, the actual attainment is somewhat greater than would be expected. We might also arbitrarily specify 130 as the average IQ of Ivy Leaguers, professors at elite colleges, and Supreme Court law clerks. Assuming an average IQ of 110 for Jews, we would expect them to be overrepresented by a factor of 4, far less than the approximately 15 to 1 that obtains. Assuming an average IQ of 115 for Jews, we would expect Jews to be overrepresented at IQ 130 by a factor of only about 7 to 1, again substantially less than the actual ratio. If we take these numbers seriously—and I actually recommend taking them with more than a grain of salt as well—then Jews are not only extraordinarily high achievers, they reach the heights in good part by overachieving.

It is important to note that the average IQ of Sephardic (mostly North African) Jews is apparently no higher than that of non-Jews and is considerably lower than that of Ashkenazi Jews. This is true even of Sephardic and Oriental Jews in Israel.

In Their Genes?

Does the very large difference in IQ and intellectual accomplishment between Ashkenazi Jews and non-Jewish Westerners require a genetic explanation? There is certainly no dearth of such explanations for the Jewish intelligence advantage. Here I mention just five of the most frequently invoked explanations.

1. *Persecution's bounty.* One very old genetic explanation is that the persecution of European Jews hit hardest at the less intelligent, who were presumably less clever at escaping the enemy. The result of trimming disproportionately the least intelligent Jews from the gene pool was to shift ever upward the average intelligence of the remaining members of the group. There are two problems with this explanation. First, it is far from clear that it was the least intelligent Jews who were most likely to be killed by pogroms. We can as convincingly argue that the most economically successful and intelligent would have been particularly conspicuous and likely to be attacked. Second, it is not clear that disproportionate elimination of the less intelligent from the gene pool would have had much effect. The phenomenon presumed by this explanation is what is known as a genetic "bottleneck"—the result of trimming some genotypes from the breeding population because of an unusual environmental circumstance. (The classic example of a bottleneck is when a small portion of a breeding population in a particular region moves away from that region, bringing with it only a limited portion of the original genetic variation.) Bottleneck effects, however, could not plausibly account for much in the way of elevated IQ for Jews, even under the assumption of very high heritability of IQ. Complete elimination from reproduction of the lowest 15 percent of the IQ distribution would raise the average IQ of the subsequent generation by about 1 point. There would have had to have been very thorough elimination of large numbers of people in the lower reaches of the gene pool, over many different occasions, to make much of a genetic difference.

2. *Nebuchadnezzar's favor.* The geneticist Cyril Darlington has proposed that the Babylonian captivity of the Jews led to their intellectual enhancement. Jerusalem fell in 586 BC, and according to the Bible, Nebuchadnezzar "carried into exile all Jerusalem: all the officers and fighting men, and all the craftsmen and artisans . . . Only the poorest people of the land were left" (2 Kings 24:10). This hypothesis further specifies that the relatively

unintelligent Jews who were left behind would have drifted into other religions, so that when the Jews returned to the Holy Land, they would not have had further interaction with the unintelligent individuals left behind. Aside from the stipulation that the unintelligent Jews drifted away from their religion, which is highly speculative, the theory rests on a hypothesis of a single bottleneck. Even under the assumption of very high heritability of intelligence and under the further, dubious, assumption that intelligence would have been highly correlated with wealth status in ancient times, the trimming off of a very substantial fraction of the poorest people from the breeding population would have left the next generation very little less intelligent than the previous generation.

3. *Marry the scholar.* Another popular genetic theory is that the daughters of (highly intelligent) merchants and businessmen were likely to be married off to the (highly intelligent) scholar or rabbi. The fruits of such unions would have been better off financially, and their offspring would have been more likely to survive. Adherents to this theory sometimes point to the injunction in the Talmud to marry a scholar. But arguments have been made that Jews both rich and poor were reluctant to entrust their daughters to penniless scholars, and the rich would have preferred a businessman. Moreover, the number of individuals involved in the presumably IQ-beneficial unions would have been a small fraction of the total population, making it dubious that they could have elevated the population average much at all.

4. *Talmud dropouts.* Political scientist Charles Murray has proposed that mere literacy would not have been sufficient to satisfy the requirement of being able to read a difficult text—the Talmud—and comprehend it, let alone provide an exegesis. Those who were unable to function at such a high level of literacy would have drifted away from the group, leaving behind only the most intelligent to propagate future generations. Murray's theory is interesting, but nothing more than speculation.

5. *Occupational pressure.* Anthropologists Gregory Cochran, Jason Hardy, and Henry Harpending have proposed by far the most elaborate theory of inherited Ashkenazi Jewish intelligence.

The Ashkenazim enter the historical record of Europe in about the ninth century. From a very early point they were engaged in occupations requiring literacy, numeracy, and generally high intelligence, including money lending (forbidden to Christians because of usury laws), trading, and—especially in Eastern Europe—tax-farming and estate management. These occupations brought wealth to those who could handle them. Wealth meant greater survival of offspring. Hence more intelligent people had more offspring than less intelligent people, and the average intelligence of the population slowly increased.

Cochran and his colleagues propose a specific mechanism underlying increased Ashkenazi intelligence. Their account begins with the fact that Ashkenazim have a particular propensity to illnesses involving storage in the nerve cells of so-called sphingolipids, which form part of the insulating outer sheaths that allow nerve cells to transmit electrical signals and encourage growth of dendrites. These illnesses include Tay-Sachs, Niemann-Pick, and Gaucher's disease. Storing too much of these sphingolipids is fatal or at least likely to result in a serious illness that often prevents reproduction.

But why should an increase in sphingolipids result in higher intelligence for the nondiseased portion of the population? At this point, Cochran's group enlists the sickle-cell anemia analogy. The sickle-cell gene produces illness for individuals who have two copies of it (one from each parent). But those with only one copy of the gene are provided with protection against malaria. This is useful for West Africans, who are threatened in their natural habitat by malaria and are by far the most likely people in the world to have the sickle-cell gene.

The proposed analogy to high intelligence for the Ashkenazim is that two copies of the extra-sphingolipid gene result in serious illness or death, but one copy of it results in high but nonlethal levels of sphingolipids. And high levels of sphingolipids encourage transmission of nerve signals and increased growth of dendrites. Presumably more expansive branching of these neuron extensions favors learning and general intelligence.

Apparently sphingolipids do facilitate nerve transmission and encourage neural branching. But otherwise the only evidence for the sphingolipid theory is that people with Gaucher's disease are highly intelligent even as compared with other Jews. Gaucher's victims in Israel have unusually high occupational status on average, and there are as many physicists among Gaucher's patients as there are skilled workers.

The sphingolipid theory modeled on the sickle-cell analogy does have one virtue over the other speculations about the genetic superiority for Ashkenazi Jews. It makes the clear prediction that people who have only one copy of the extra-sphingolipid gene should have higher IQs than those who have no copy of the gene. However, Cochran and colleagues did not test this prediction, other than reporting impressive occupational attainment for Gaucher's victims. I find this puzzling. The hypothesis is not hard to test, and many scientists would have tested this prediction before publishing a complicated theory.

Note that the Cochran theory has an unusual property in that it attributes no great causal significance to the fact that the Jews had universal literacy at such an early stage in history. Literacy was important only insofar as it made it easier for the Jews to engage in the occupations that ultimately produced high intelligence for the group. Indeed, it is important to the Cochran theory that there not be much Jewish intellectual accomplishment prior to the time that occupational advantages began to work their genetic magic. And it is certainly true that intellectual accomplishments of a high order by Ashkenazi Jews came with ever greater frequency over the centuries, reaching a crescendo by the middle of the nineteenth century.

It is also important to the Cochran theory that Sephardic Jews not be terribly accomplished, since they did not pass through the genetic filter of occupations that demanded high intelligence. Contemporary Sephardic Jews in fact do not seem to have unusually high IQs. But Sephardic Jews under Islam achieved at very high levels. Fifteen percent of all scientists in the period AD 1150–1300 were Jewish—far out of proportion to their presence in the world population, or even the population of the Islamic world—and these scientists were overwhelmingly Sephardic. Cochran and company are left with only a cultural explanation of this Sephardic efflorescence, and it is not congenial to their genetic theory of Jewish intelligence.

In short, there are many genetic theories of Jewish intelligence but not much in the way of convincing evidence.

Other Cultures with High Levels of Intellectual Accomplishment

Jewish intellectual eminence needs to be put into context with the fact that intellectual achievements have often differed hugely between cultures—even between cultures with high rates of literacy and robust economies. And it is impossible to give a genetic account to these fluctuating rates of accomplishment.

In AD 1000, the intelligentsia of the world consisted primarily of Arabs and Chinese. Arab sheiks were discussing Plato and Aristotle, and Chinese mandarins were practicing all the arts, at a time when European nobles were gnawing haunches of meat in cold, dank castles. The intellectual scoreboard was lit up for Chinese and Arabs (and Indians) when the tally was near zero for Europe. Then Europe gradually came to intellectual preeminence, partly because of its willingness to learn from the more advanced cultures. Changes in the gene pool are an impossible explanation for this enormous shift in the intellectual center of gravity.

Even within Europe the swings of intellectual prominence have been extraordinarily wide. Spain was at the height of intellectual achievement under the Moors, but sank rapidly thereafter, never achieving much of great note even in the heady days of New World gold and silver. Northern Italy was a powerhouse in all the arts and sciences in the fifteenth century, a time when England was a cultural backwater. Since 1800, England has been a leader in almost all realms of endeavor and a lion in science, philosophy, and literature. Italy since 1800 has been a shadow of its Roman and Renaissance self. The Scots, well past the late Middle Ages, were savages painting themselves blue for battles and frequently choosing leaders by means of assassination. (Shakespeare had good reason to site *Macbeth* in Scotland.) By the eighteenth century, the Scots were leaders in science and philosophy. Scandinavia was not noted for intellectual achievements until the twentieth century.

Within the United States, the regional differences in intellectual accomplishment have been nothing shy of astonishing. The Northeast has never had much more population than the South, but easterners have achieved incomparably more in science, philosophy, and the arts (save for music) than have southerners. The population of Texas has exceeded that of New England for the last hundred years by a factor of 3 or 4; even the white, non-Hispanic population of Texas is larger than that of New England. But a visit to *Who's Who* establishes that Texans have achieved virtually nothing in the sciences or philosophy (though there have been notable achievements in literature, music, and the arts in recent decades).

The magnitude of the differences in intellectual accomplishment between Jews and non-Jews in the West pales beside all these national, ethnic, and regional differences. This still leaves us, however, with the differences in IQ between Jews and non-Jews. Can this plausibly be accounted for exclusively by cultural factors?

Cultural Explanations of Jewish Intelligence

Education has been as important to the Jewish tradition as to the Confucian tradition. As Austrian novelist Stefan Zweig wrote during World War II, "It is counted a title of honour for the entire family to have someone in their midst, a professor, a savant, or a musician, who plays a role in the intellectual world, as if through his achievements he ennobled them all."

The analogy between Jews and Confucians is marked in another crucial respect as well. Family ties are very strong for Jews, and the family can make demands on the individual that are hard to resist. The Jewish mother, fabled in song and story, is reputed to have an influence that seems equal to that of any Chinese patriarch. And much of her influence is directed toward educational and intellectual attainment.

The Jewish value of education takes some forms that seem extreme to non-Jews. Psychologist Seymour Sarason was raised in a poor neighborhood, and his family never had money. But Sarason reports that when one of his cousins decided to play football in high school, the family was angry and disbelieving. If he got injured, the family worried, he might not be able to go to college. (Compare with *Friday Night Lights*.) Sarason also recounts that his father, who had little money and less education, once bought an expensive encyclopedic dictionary for the family. This made a huge impression on Sarason. It told him that education was important—even at great financial sacrifice.

In Brooklyn today, while other boys trade baseball cards, Hasidic boys trade rabbi cards.

And so on and so forth. It would be possible to list numberless anecdotes showing that Jews value intelligence, the intellectual life, and achievement. The value on achievement is certainly not limited to intellectual fields. In addition to business success, Jews in the past have valued athletic success. In the early part of the past century, high achievement included preeminence in boxing, wrestling, and basketball (which one anti-Semitic commentator claimed was the quintessential Jewish sport—because it involved sneakily stealing the ball). So it seems that Jews place a high value on achievement, period.

But the sum of evidence about the cultural value placed on intellectual achievement would amount to only a pile of anecdotes. We have no quantifiable measure of cultural influences on intellectual achievement, not even systematic anthropological observations of the kind that Shirley Brice Heath made comparing middle-class whites, working-class whites, and poor blacks.

We can, however, speculate that culture has resulted in an increase for Jews in phenotypic IQ. (This is the IQ that the individual has as a consequence of the environment operating on the individual's genetic makeup in a particular way.) Recall that the children of the Chinese American members of the class of 1966, who were themselves of average or slightly less than average IQ but attained very high occupational status, had IQs averaging 109 when they were quite young. That was before these children encountered the U.S. public school system and engaged with the non-Asians in their neighborhoods. By late childhood the average IQ of the children of the class of '66 had dropped to 103. But the grandchildren of the class of '66 would have an advantage even over the children of the class of '66 because the children of the class of '66 had higher (phenotypic) IQs than their parents and thus would likely have created yet more advantaged environments. The consequence is that the IQ of the grandchildren of the Asian American class of '66 might well be greater than 103.

The Asian American example of one generation building IQ on the socioeconomically advanced shoulders of the preceding generation shows that culture could account for a significant portion of the IQ gap between Jews and non-Jews—perhaps even all of it—by scaffolding to ever higher levels.

In any event, Jewish intellectual achievement is probably at least in part overachievement. Achievement at such ultrahigh levels seems substantially greater than would be predicted by the IQ advantage.

CHAPTER TEN

Raising Your Child's Intelligence . . . and Your Own

IN THIS CHAPTER I remind you of some things you already know about how to increase the intelligence of your child and yourself, point out some things that you may believe to be true but for which there is little evidence, and describe some ways to increase intelligence that may be surprising to you. By *intelligence,* I mean ability to solve problems and to reason, which is measured, if only imperfectly, by IQ scores and academic achievement.

The Obvious

First, the things you probably do instinctively with your child, without much conscious consideration of strategy, can increase intelligence: Talk to your child, using high-level vocabulary. Include your child in adult conversations. Read to your child. Minimize reprimands and maximize comments that will encourage your child to explore the environment. Avoid undue stress, which you would do anyway, but probably not for the reasons given in this book. Stress can result in poor learning ability and ability to solve novel problems as it can damage pathways between

the limbic lobe and the prefrontal cortex. At extremes, stre
interfere with memory capacity as well.

Teach your child how to categorize objects and events and h
to make comparisons among them. Encourage your child to analyze and evaluate interesting aspects of the world. Give your child
intellectually stimulating after-school and summertime activities.
(Though I have to admit that I do think some parents overpro-
gram their kids. The forced suburban march from hockey practice
to piano class to Cub Scout meetings is not something I would
personally recommend.) Try to steer your child toward peers who
will promote intellectual interests.

These are the sorts of things that people of higher socioeconomic status (SES) are more likely to do than people of lower SES,
and they are all correlated with children's ultimate intelligence.
Admittedly though, the data are just that—correlational—for the
most part. We do not know the extent to which activities like
these cause greater intelligence as opposed to just being the sorts
of activities that smarter parents carry out with their children—
who are destined to be smarter than average because of their
parents' good genes and not their exemplary behavior. On the
other hand, common sense indicates that these things ought to be
beneficial and it's hard to see how they could do harm. And we do
know that when poor children are raised by higher-SES parents,
who are inclined to do these things, the children end up having
higher IQs and better academic performance.

The Dubious

Despite what you may have heard bandied about in the press,
some things do not affect intelligence much—or at any rate there
is not very good evidence that they do. Baby Einstein educational
toys that move around and communicate with the child may be as
likely to induce passivity as to encourage exploration. There is no
evidence that playing Mozart to your child—whether born yet or

not—will increase intelligence. The research suggesting that extra stimulation in the early years results in more growth of neurons and better problem-solving ability is based strictly on animal studies comparing rats that had absolute minimal stimulation—sitting in the dark in small cages—with animals that were allowed to play in interesting environments and given a companion to do it with. We find the same kind of huge gains with infants who are drastically understimulated and then brought into normal environments. We do not know whether stimulation at unusually high levels using fancy toys does much for human infants.

But lots of other things do seem to matter, and for many of them we have very good evidence about their usefulness.

The Physical

Though pregnant mothers may worry about whether exercise is risky, there is evidence that exercise increases the intelligence of the newborn without risk to mother or child. In an exceptionally well-designed study of forty pregnant women, all of whom were accustomed to exercising frequently and strenuously, investigators asked half of the women to exercise vigorously at least three times per week—to run, perform aerobics, or ski cross country. The investigators asked the other half of the women to limit their exercise to walking. The babies born to the exercising mothers had IQs nine points higher at age five. The study needs replicating, however, because the nine-point difference seems improbably great.

Exercise is good for the baby, for mothers-to-be, and for everybody else. Exercising large muscle groups actually increases growth of neurons, and exercise, at least in animals, adds to the blood supply of the brain. Even introducing exercise relatively late in life is good for intelligence. Experiments show that elderly people who are encouraged to exercise maintain good problem-solving skills longer than people who are not encouraged to exercise. The effect of exercising thirty minutes or more per day on fluid intelligence–based tasks is .50 SD across all studies. Strength training plus cardiac training is better than cardiac training alone. People who exercise regularly in middle age are one-third as likely to get Alzheimer's disease in their seventies as people who do not exercise. You can even start exercise in your sixties and reduce the likelihood of Alzheimer's by half.

A possibly very important thing a mother can do for her child's intelligence is to breast-feed. For children with the most common kinds of genetic makeup, breast-feeding for up to nine months may increase IQ as much as 6 points. (Breast-feeding beyond nine months seems to have no beneficial effect.) It seems to be particularly important to breast-feed premature babies.

Fluid-Intelligence Exercises

Several types of activities can improve fluid intelligence, and not just for children. Recall that fluid intelligence is the ability to solve problems that are novel and for which previously learned rules or concepts are not necessarily helpful. The prototypical example is the Raven Progressive Matrices test. You see various geometric figures that have been changed in particular ways and you have to develop on the spot a rule that will allow you to determine what the next transformation of the figures should be. The activities that increase fluid intelligence include computer games that teach attention control and exercise working-memory capacity.

Neuroscientist Rosario Rueda and her colleagues described several types of games that exercise fluid-intelligence functions for young children. One was *anticipation* exercises. An exercise they used for this involved teaching children to anticipate where a duck that had submerged itself in a pond would emerge. The children used a joystick to manipulate a cat to the position where they expected the duck to turn up. Another exercise involved *stimulus discrimination*. Children had to remember the attributes of cartoon portraits so that they could pick the portrait out of an array of several other portraits. Other tasks, described in Chapter 3, included a *conflict-resolution* exercise and an *inhibitory-control* exercise. These exercises improved performance on the Raven Progressive Matrices test, which makes heavy demands on attention control and working memory. These fluid-intelligence functions are particularly important for children's learning in the years before adolescence. For a demonstration of these exercises, go to http://www.teach-the-brain.org/learn/attention/index.htm

Child neurologist Torkel Klingberg and his coworkers found that various working-memory and attention-control tasks improved the concentration of children with attention deficit hyperactivity disorder (ADHD). Some of these tasks are easy to duplicate without using a computer. For example, experimenters read to children a series of digits (perhaps 4, 7, 2, 9, 5) and then ask them to repeat the digits back in the reverse order in which they were read. Other tasks require computer administration, such as a "go–no go" task in which the child is shown two gray circles and has to press a key if a circle turns green and refrain from pressing it when it becomes red. Such exercises improved reaction times and reduced errors even on tasks for which the children had not been trained, and improved scores on the Raven IQ test. The Raven scores for almost all the trained children exceeded the average for the untrained children. Similar exercises improved working memory and Raven performance in normal, healthy adults.

Finally, meditation exercises typical of the kind used in Chinese traditional medicine (breathing exercises, postural training, body awareness) conducted over a period of just five days improved executive functioning and performance on Raven Progressive Matrices. If this seems dubious to you, it would to me too, except that the authors of the study include highly respected neuroscientists Yi-Yuan Tang and Michael Posner, so I believe it.

Self-Control

The best evidence we have indicates that children with above-average self-control have higher intelligence, and higher academic achievement whatever their level of intelligence. Personality psychologist Walter Mischel and his colleagues found that the largely upper-middle-class children at Stanford University's nursery school who could delay gratification in the here and now (one cookie) for better rewards later (two cookies) got higher grades and substantially higher SAT scores when they were teenagers. Lower-SES minority children in New York also got higher grades if they had greater ability to delay gratification. We do not know, however, whether the ability to delay gratification is associated with later scores on ability tests merely because children with more intelligence at age four incidentally happen to have better ability to delay gratification, and then become smarter teenagers not because their ability to delay gratification helped them to learn but because they were destined to be smarter owing to other inherited or environmental reasons. But it does seem likely that the ability to delay gratification in itself increases ability because greater self-control makes studying easier. Recall that psychologists Angela Duckworth and Martin Seligman found that junior high school students in a magnet school in Philadelphia who had greater self-control had higher grade point averages. In fact, the correlation between self-control and grade point average was twice as great as the correlation between IQ and grade point average. Here we are on safer causal ground. Self-control almost surely contributes to achievement over and above the intelligence level that a person happens to have.

Unfortunately, we are not confident about knowing ways to increase self-control in children, but the research provides some hints. We do know that if children watch adults who reward themselves regardless of their performance, they are more likely to do the same for themselves. But if they watch adults reward themselves only for high-quality performance, children do that themselves. Also, Mischel and his coworkers had a few tricks that helped the children in their study to delay consuming a goodie immediately as opposed to waiting for a bigger reward.

When researchers had the children "think fun thoughts" instead of thinking about the rewards, the children were able to delay longer. When researchers encouraged them to put the rewards away, out of their line of sight, they also delayed longer. We do not know whether these kinds of suggestions would generalize to behavior outside the lab session where they were taught, but they might. And if parents were to look for occasions to encourage children to be patient, and especially if they gave suggestions about how to be patient, this might be effective. Parents might also try modeling delay of gratification. Mischel's group found that children behaved like the adults they watched. Some children saw an adult take an immediate reward instead of waiting for a larger one later. The adult said things like, "You probably have noticed that I am a person who likes things now. One can spend so much time in life waiting that one never gets around to really living." Even children who were inclined to delay gratification, but who watched such a model, subsequently took the immediate reward most of the times they were offered it.

Teach Malleability—and Praise Children for Hard Work

It is crucial for parents to teach children that their intelligence is under their control. Asians are particularly likely to believe that ability is something you have to work for. Not surprisingly, Asian Americans work harder to achieve academic goals than European Americans. And Asians work harder after failure than after success—unlike North Americans of European descent who work harder after success than after failure. It is important to teach children that if at first you don't succeed, try again harder.

It is probably a bad idea to praise children for being intelligent. Instead, praise hard work, which is under their direct control. The problem with praising children for their intelligence is that it makes them focus on trying to show how smart they are by working on tasks they do well on and avoiding working on tasks they are having trouble with. When children are praised for intelligence, in other words, they resist accepting a challenge and doing things from which they can learn a lot.

In a clever experiment illustrating this point, developmental psychologists Claudia Mueller and Carol Dweck told children that they had done very well on problems from the Raven Progressive Matrices test and praised them either for being bright or for working hard. They then offered the children the opportunity to work on another set of problems—either easy ones ("so I'll do well") or hard problems that would challenge them ("so I'll learn a lot from them, even if I won't look so smart"). Sixty-six percent of the children who were praised for their intelligence chose to work on easy problems that would show that they were smart; over 90 percent of children praised for hard work chose problems that they would learn a lot from. If the children did well because they were smart, they did not want to risk finding out that they were not so smart after all. If they did well because they worked hard, they wanted problems that would test their limits and teach them how to do even better.

Before the children actually got a chance to work on a problem set of their choice, Mueller and Dweck required them to work on a second set of problems that were much more difficult than the first set. The children were then asked to explain why they had performed poorly on the second set of problems. The children praised for intelligence based on performance on the first set of problems were more likely to think that their failure on the second set of problems reflected lack of ability; children praised for hard work initially were more likely to think that their failure on the second set of problems was due to lack of effort. Children praised for ability were less likely to want to continue to work on the problems and reported enjoying working on the second set of tasks less than did those praised for hard work. As icing on the cake, Mueller and Dweck then had the children work on a third set of problems. Children who had initially been praised for intelligence solved fewer problems than those initially praised for hard work. The moral of this experimental parable seems clear: praise for effort, not smarts.

Avoid "Contracts" to Give Rewards for Activities That Are Intrinsically Rewarding

It is not a great idea to promise your child a reward for doing something you want to encourage, if your child already has some interest in it. With developmental psychologists Mark Lepper and David Green, I watched nursery school children engage in a novel activity—drawing with magic markers. Most of the children drew with the markers and clearly enjoyed the activity. We later promised some children a reward if they would draw something with the magic markers for us, which they gladly did. Then a couple of weeks later the magic markers were put out for children to play with again. Children who had been rewarded for playing with the magic markers drew with them less than children who had not been rewarded—and their drawings were of lower quality. In effect, the "contract" had turned play into work. We praised the products of other children, who were not promised a reward, and these children subsequently played with the magic markers more than did children who were neither promised a reward nor praised. So if you want children to do something, praise them for doing it. Don't promise them a reward for doing it.

Sometimes, however, contracting for rewards can be a good idea. If the child is not going to do something without being offered an extrinsic reward, then rewards may have to be the order of the day. If a child has low initial interest in an activity, the reward may serve to get the child to try it and perhaps find that there are genuine attractions to it. I suspect that the reward aspect of the KIPP charter schools may be a good idea for their children, many of whom will have found little to interest them in their previous schools.

Raising Your Child's Intelligence . . . and Your Own   191

Effective Tutoring

When you tutor your children, try to keep in mind Mark Lepper's five Cs tutoring guide from Chapter 4: encourage a sense of *control, challenge* your child, instill *confidence,* foster *curiosity,* and *contextualize* by relating the task to the real world or to a movie or TV show. In addition, don't sweat the small errors like forgetting to write down the minus sign; try to prevent the child from making a mistake unless there is a good lesson to be learned from it; don't dumb down the material for the sake of the child's self-esteem but rather change the way it is presented; ask leading questions; and *don't* give much praise so as to avoid making the child feel evaluated.

The Schools

Finally—some suggestions about dealing with the schools. To the extent you can, try to get your child into classrooms with the best teachers, especially for the first grade. Avoid rookie teachers. If your school does not use proven computer programs for teaching reading, math, and science, try to get it to consider doing so. Go to the U.S. Department of Education's Web site for What Works Clearinghouse so that you can cite chapter and verse for why certain programs should be used at your child's grade level. If your child's school doesn't use any of the cooperative learning tools, where children work on solving problems and creating knowledge together, encourage the school to do so, again citing the What Works Clearinghouse. Find out if the principal at your child's school is aware of who the good teachers are and ask if it is possible to reward the better ones. If it's not possible to reward teacher quality, press your school board to make it possible. (Union contracts may forbid rewarding on the basis of anything but seniority. In that case, you can encourage your school board to reward all teachers at high-performing schools.) Discourage your school board from putting a lot of emphasis on teachers getting certification and higher degrees, because there is no evidence that teachers with certificates or higher degrees are any better at their jobs. Teacher time is better spent working on teaching skills, with the help of peers and experts who observe them and give them feedback.

In short, you can use many of the lessons in this book to improve your children's intelligence—and your own.

What We Now Know about Intelligence and Academic Achievement

THE STRONG HEREDITARIAN view of intelligence holds that intelligence is mostly a matter of genes. You are only going to be as smart as your genes allow, and nothing much in the environment—neither the way you are brought up in your family nor the kind of schools you go to—is going to significantly change that. Many if not most scientists who study intelligence and many if not most laypeople in the United States believe this. Such thinking is extremely unfortunate for the individual, because it implies that hard work can produce little in the way of improvement to "real" intelligence. It is a disaster for public policy because it implies that educational interventions are doomed to failure. Fortunately, the view is quite wrong. And here is how we know.

There is no fixed value for the heritability of intelligence. It differs from one population living in one set of circumstances, to another population in another set of circumstances. If the environment is highly favorable to the growth of intelligence, then the heritability of intelligence is indeed fairly high—perhaps as high as 70 percent. This is the situation that exists for the upper middle class in developed countries. The environments they create promote intelligence, and one family differs little from another in that respect. At the limiting extreme of identical environments for everyone in a given group, the only factor that can influence differences in intelligence is genetics. The upper middle class comes close enough to that situation that heredity for that group can be very important in determining differences in intelligence.

But if the environment is highly variable—differing greatly between individual families—then the environment is going to play the major role in differences in intelligence between individuals. And this is the situation for the poor. For them, only 10 percent of the variation in intelligence is driven by heredity, which means that improving the environment of children born into poor families could have a big effect on intelligence. And in fact if you raise a poor child in an upper-middle-class household, the expected value added is at least 12 IQ points and may be as high as 18 IQ points. The effect on academic achievement is also very great—at least half a standard deviation and under some circumstances as much as a full standard deviation.

Entirely aside from the degree to which heritability is important for one group or another in the population, heritability places no limits whatsoever on modifiability—for anybody. The height of people in developed countries has increased greatly in recent generations—and this increase has nothing to do with genetics.

There has been a similarly striking increase in IQ over the past century. The scores on IQ tests have increased by more than 18 points in the last sixty years and probably by the better part of two standard deviations (30 points) over the past hundred years. And on the Raven Progressive Matrices—the test heralded for decades as being a culture-free measure of true intelligence—the gain has been two standard deviations in less than sixty years.

Why have the gains occurred? It is simple at base: the schools and the culture have changed radically in such a way as to affect scores on many of the subtests of IQ tests. Parents and schools increasingly teach children how to categorize objects and events in taxonomic terms suitable for scientific analysis. The media teach children the ways of the world—why policemen wear uniforms, why street addresses are numbered in order, and why people pay taxes—resulting in higher scores on comprehension subtests of IQ tests. Improvements on the Raven matrices—and in the fluid intelligence underlying performance on it—can be traced at least in part to the ever-more geometric and analytic ways of teaching arithmetic over recent decades, and possibly in part to computer games. A few years ago McDonald's was including in its Happy Meals mazes that were more difficult than the mazes in an IQ test for gifted children!

And then there is the fact that people are receiving a lot more education than ever before. In a century the mean number of years of schooling has gone from seven to fourteen. Since a year of school adds as much to IQ scores as two years of age, it would be astonishing if IQ had not changed radically over that period.

How much of the IQ gains can be called "real" intelligence gains? I can say several things about that. First of all, it is out of the question that the great-grandchildren of people who were ten years old in 1910 are two standard deviations smarter, if we define intelligence broadly as the "ability to reason, plan, solve problems, think abstractly, comprehend complex ideas . . . and learn from experience." On the other hand, people today really are more intellectually capable than their forebears. Children who can tell you why street addresses are numbered consecutively are smarter in some sense than children who can't. Being able to think about the similarities of objects and events in taxonomic terms is a real advantage. Heuristics for reasoning, such as procedures for hypothesis testing, are part of the curriculum at every level of school, and they can be applied to everyday problems. Planning and choosing are two aspects of intelligence that have been increased by virtue of widespread knowledge of probability theory and cost-benefit reasoning.

Since school makes children smarter, there is no doubt that better schools can make them smarter still. Although vouchers, charter schools, whole-school interventions, and teacher certification or higher academic degrees do not reliably improve education, other factors do—and some matter a great deal. Teachers differ a lot in quality, and so finding ways to improve the quality of teaching could make a great difference. If we could replace the bottom 5 percent of teachers every year with average-quality teachers, the level of children's academic performance would increase hugely in just a few years. Use of computer-assisted forms of teaching can produce huge gains in the rate of learning, and some types of cooperative learning are highly effective. And recall the Herrnstein demonstration with an intensive program in Venezuela that radically improved the problem-solving skills of ordinary junior high school students. It also raised their IQ scores by a nontrivial amount—5 points on a typical test of multiple problem-solving skills.

The received opinion about the relationship between social class and intelligence is that intelligence, which is largely inherited, drives social class. Smarter people have better genes so they are destined to rise in society, whereas less smart people have worse genes so they are destined to fall. It is true that intelligence is partially heritable, and more intelligent people on average will be of a higher social class in virtue of their greater inherited intelligence. But I believe that the role of genetic inheritance in determining social class is fairly small. The difference between the average IQ of the children of the lower third of the socioeconomic status (SES) distribution and the average IQ of the children of the upper third is about 10 points. We know that some of this is due to biological but not genetic factors, including exercise, breast-feeding, and exposure to alcohol or cigarette smoke, as well as hazardous chemicals and pollution. And some of it is due to the disruption in schools of lower-SES children and to the fact that peers are pulling intelligence mostly in a down direction. We also know that socialization in lower-SES homes is not optimal for developing either IQ or school readiness. Moreover, a child born into roughly the bottom sixth of the SES distribution will have an IQ 12 to 18 points higher if raised by parents from roughly the top quarter of the SES distribution. All of this does not leave much room for genes in the social-class equation. I do not doubt that genes play a role, but I would be surprised to find that the differences in inherent genetic potential of the social classes are very great. Certainly much if not most of the 10 points separating the average of the children of the lower third and the average of the children of the upper third is environmental in origin.

For the race difference in IQ, we can be confident that genes play no role at all. Most of the evidence offered for a genetic component to the race difference is indirect and readily refuted. Virtually all of the direct evidence, which is due mostly to the natural experiment resulting from the fact that American "blacks" range from being completely African to largely European in heritage, indicates no genetic difference at all with respect to IQ. And the difference between the races in both IQ and academic achievement is being reduced at the rate of about one-third of a standard deviation per generation. The IQ of the average black is now greater than that of the average white in 1950.

The No Child Left Behind Act demands that the difference in academic achievement between the classes and between the races be erased in half a generation by the schools alone. This is absurd. It ignores the fact that class and race differences begin in early infancy and have as much to do with economic factors and neighborhood and cultural differences as with schools.

That is the bad news about gap reduction. The good news is that big improvements in IQ and academic achievement for lower-SES and minority children are possible. And we know at least the outlines of what those improvements look like. Half-measures have been tried and are not going to make a lot of difference. We need intensive early childhood education for the poor, and we need home visitation to teach parents how to encourage intellectual development. Such efforts can produce huge immediate gains in IQ and enormous long-term gains in academic achievement and occupational attainment. Highly ambitious elementary, junior high, and high school programs can also produce massive gains in academic achievement. And a variety of simple, cost-free interventions, including, most notably, simply convincing students that their intelligence is under their control to a substantial extent, can make a big difference to academic achievement.

Believing that intelligence is under your control—and having parents who demand achievement—can do wonders. At any rate that has been true for Asians and Jews. There is no reliable evidence of a genetic difference in intelligence between people of East Asian descent and people of European descent. In fact, there is little difference in intelligence between the two groups as measured by IQ tests. Some evidence indicates that East Asians start school with lower IQs than do white Americans. After a few years of school this difference seems to disappear. But the academic achievement of East Asians—especially in math and the sciences, where effort counts for a lot—is light-years beyond that of European Americans. Americans of East Asian extraction also differ little in IQ from European Americans. In any case, the academic achievement and occupational attainment of Asian Americans exceed by a great amount what they "should" be accomplishing given their IQs. The explanation for the Asian/Western gap lies in hard work and persistence.

Jewish culture undoubtedly has similarly beneficial effects. Jewish values emphasize accomplishment in general and intellectual attainment in particular. Differences between Jews and non-Jews in intellectual accomplishment at the highest levels are very great. A genetic explanation for this is not required inasmuch as even greater differences have occurred for Arabs and Chinese versus Europeans in the Middle Ages, for differences between European countries at various points since the Middle Ages (with reversals occurring between Italy and England and with movement from savagery to sagacity in scarcely two centuries in Scotland), and for regional differences in the United States. We are left with an IQ difference of two-thirds to a standard deviation between Jews and non-Jews. At least some of this difference is surely cultural in origin.

Finally, there is much that we can do to increase the intelligence and academic achievement of ourselves and our children. Everything from the biological (exercise and avoidance of smoking and drinking for pregnant women, and breast-feeding for newborns) to the didactic (teaching categorization, following good tutoring principles) can make a difference to intelligence.

We can now shake off the yoke of hereditarianism in all of our thinking about intelligence. Believing that our intelligence is substantially under our control won't make us smart by itself. But it's a good start.

APPENDIX A

Informal Definitions of Statistical Terms

ALL KINDS OF PHENOMENA are found to be distributed normally, that is, in the shape of a *bell curve*, shown in Figure A.1. For example, if we were to plot on a graph the number of eggs produced weekly by different hens, the number of errors occurring during production of a given type of automobile, or the IQ test scores for a group of people, the shape of the curve representing the data would approximate that of the bell curve. We do not need to go into the mathematical reasons for why distributions tend to have this shape. What is important is that the normal distribution curve has useful properties for making inferences about where observations stand in relation to other observations. The normal curve in Figure A.1 is divided into *standard deviations*—so named because the average score deviates from the mean by (approximately) this amount. In a perfectly normal distribution, which is a mathematical abstraction but one that is approximated surprisingly often if there are a very large number of observations, about 68 percent of all observations fall between +1 and −1 standard deviation (abbreviated SD) from the mean (set at 0 in the curve in Figure A.1). Another set of useful facts about the concept of standard deviation concerns the relation between percentiles and standard deviations. About 84 percent of all observations occur at or below 1 SD above the mean; an observation at exactly 1 SD above the mean is at the 84th percentile of the distribution. Sixteen percent, the remaining observations, occur above that standard deviation. Almost 98 percent of all observations lie below 2 SDs above the mean. A score at exactly 2+ SDs from the mean is at the 98th percentile. The remaining 2 percent of observations are above that. Nearly all observations fall between 3 SDs below the mean and 3 SDs above it. By convention, the SD of the distribution of scores on most IQ tests is forced to be 15 (with a mean of 100).

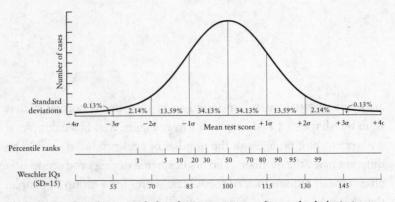

Figure A.1. The normal distribution curve, with standard deviations from the mean marked by vertical lines and with corresponding percentile scores and Wechsler IQ scores given below. Note that 68 percent of values fall between 1 standard deviation (σ) below the mean and 1 standard deviation above the mean.

Standard deviations are useful units with which to describe *effect sizes*, for example, in determining how much difference a new teaching technique makes to what the students learn. The most common indicator of effect size is a statistic called Cohen's *d*, which is calculated as follows: the mean of group A minus the mean of group B divided by the average of the standard deviations of the two groups (or sometimes divided by the standard deviation just of group A).

By convention, *d* values of .20 or less are deemed small. This is the equivalent of moving the experimental group's scores from the 50th percentile to almost the 60th percentile. You might not think that was such a small effect if the scores in question were those that could be expected by your child if he or she is taught by the new technique (60th percentile) versus the old technique (50th percentile). And whether you would be willing to pay for the technique depends in part on how important the difference between the 50th percentile and 60th percentile is. If you are measuring teaching effectiveness in terms of how quickly a child learns the touch typing system to a proficiency of 40 words per minute, and the difference between the 50th and 60th percentiles amounts to a few days, you probably would not be willing to pay too much for that gain and would not want your school system to pay much for it either. If you are comparing the effectiveness of two high school math teaching techniques by looking at average scores on the SAT exam, and one technique results in an average score of 500 and the other results in an average score of 520, this is the difference between the 50th and the 60th percentiles (assuming that the SD of SAT scores is 100). You might be willing to pay some significant amount of money for it. And you might be happy if your school board paid a modest amount per pupil to spring for the more effective method.

By convention, *d* values of .50 or so are considered moderate. In the world of IQ tests and academic achievement, though, an effect size of that magnitude would normally be considered a bombshell. It is the difference between an SAT score on the math section of 500 and one of 550—sometimes enough to make the difference between being accepted to a fairly good university and a significantly better university. You and your school system might be willing to pay a considerable amount to adopt a new method that would move the average child from the 50th percentile of SAT math scores to about the 70th percentile (which is what .50 SD corresponds to).

Effect sizes in the range of .70 to 1.00 SD are considered large. For education or intelligence differences, an effect size of 1.00 is huge. The putative IQ difference between blacks and whites is on the order of 1.00 SD. In Chapter 6, I discussed whether that is the actual magnitude of the difference. If it were, it would mean that the average IQ for blacks stood at the 16th percentile of the distribution of IQs for whites. An intervention that took children from, on average, the 50th percentile of the national distribution in math achievement scores to the 84th percentile would be considered worth it even at very great cost. For a nation, the increase in competitiveness that could result from such an improvement in math scores would be worth an enormous economic outlay.

The *correlation coefficient* is a measure of the degree of linear association between two variables. For example, the correlation between IQ scores and academic grades happens to fall around .50, indicating a moderately high degree of association. At least a moderate association should be expected because IQ tests were invented to predict how well people would do in school. Correlation coefficients range between −1, indicating a perfect negative association, and +1, indicating a perfect positive association. A correlation coefficient of 0 indicates no association at all. The correlation coefficient is another measure of effect size or, rather, of relationship magnitude, with values lower than .30 considered to be small, values of .30 to .50 considered moderate, and values above .50 considered to be large. But, as with effect size, whether a correlation is important or not has more to do with the variables represented in the correlation than with the size of the correlation. The correlation coefficient is interpretable in standard deviation terms. A correlation of .25 between two variables indicates that an increase of 1 SD in the first variable is associated with an increase of .25 SD in the second variable; a correlation of .50 is associated with an increase of .50 SD. So if the correlation between class size and student achievement on a standardized test were −.25, then a decrease of class size by 1 SD could be expected to produce an improvement in test scores of .25 SD (assuming that the relationship between class size and test scores is genuinely a causal one).

Multiple regression is a way of simultaneously correlating a number of independent or predictor variables with some target or outcome variable. For example, we might want to compare the degree to which a number of variables predict the appeal of houses on the real estate market. We might measure the area in square feet, the number of bedrooms, the opulence of the master bathroom (for example, using an index based on the number of sinks, the presence or absence of a hot tub, and the use of high- or low-quality materials), the average income in the neighborhood, and the charm of the house as rated by a panel of potential buyers. We then correlate each of these variables simultaneously with the appeal of the house as measured by the amount that it can fetch on the market—the target variable. We get an estimate of the contribution of a given variable to the market value by finding out the size of the correlation of the variable with market value *net* of the contribution of all the other variables (that is, holding all other variables constant). Thus, charm, holding all other variables constant, might be correlated .25 with market value, and master bath opulence might be correlated .10 with market value. But all of the variables are going to be correlated with one another, and some of the variables are measured with greater precision than others, and some of the variables have a causal relation with others while others do not, and some variables that have not been measured are going to exert an effect on some of those that have been measured. The result is that multiple-regression analysis can mislead us. The actual magnitude of the contribution of charm to market price could be substantially higher or lower than the figure of .25 derived from the regression analysis.

There are endless numbers of instances where multiple-regression analysis gives one impression about causality and actual experiments, which are nearly always greatly preferable from the standpoint of causal inference, give another. For example, about fifteen years ago, I attended a consensus development conference put on by the National Institutes of Health. The purpose of the conference was to review research on medical procedures versus surgical procedures as treatments for coronary artery blockage and reach consensus about the appropriateness of each. The results of a very large number of expensive American studies, paid for by government funds, were available. In those studies, researchers put a host of variables about patients such as illness history, age, and socioeconomic status (SES) into a multiple-regression equation and drew conclusions about the effect of treatment type "net" of all the other ways in which patients varied. But because Internal Review Boards governing research policy in the United States require allowing patients to choose their treatment (it is far from clear that this is actually in the patients' interests), all the U.S. evidence was undermined by the *self-selection* artifact (see below). But in addition to the American studies were two European studies based on the random assignment of patients to treatment. Quite correctly, the panelists ignored the expensive American studies and considered only the results of the European studies.

Let's consider an example closer to the topic of this book, namely, whether size of class matters to student performance. Multiple-regression analysis tells us that, net of size of school, and average income of families in the neighborhood where the school is located, and teacher salary, and percentage of teachers who are certified, and amount of money spent per pupil in the district, and so on, average class size is uncorrelated with student performance (Hanushek, 1986; Hoxby, 2000; Jencks et al., 1972). But one well-conducted, randomized experiment varying class size over a substantial amount (13 to 17 pupils per class as compared to 22 to 25 pupils per class) found that class size varied to that degree produces an improvement in standardized test performance of more than .25 SD—and the effect on black children was greater than the effect on white children (Krueger, 1999). This was not merely another study on the effects of class size. It *replaced* all the multiple-regression studies on class size.

I occasionally cite multiple-regression studies in this book, but sparingly and always with a warning to beware the results.

Self-selection is one of the problems underlying the difficulty of interpreting correlational studies and multiple-regression analyses, and it is crucial to understand, for many reasons. When we say that IQ is correlated with occupational success to a particular degree—say, .40—there is a reflexive tendency to assume that the relationship is entirely causal: higher IQ makes a person perform a job better. But IQ is correlated with other factors too. For example, higher IQ in a child is associated with higher SES of that child's parents, which, for example, makes it more likely that the child will go to college regardless of the child's IQ level. And a college education, again regardless of IQ level, makes higher occupational status more likely. Thus, the correlation between IQ and occupational success is contaminated by the contribution of other variables like parents' SES and college attendance, which the child, or subject, has been allowed to "self-select." (It is odd to say that a person "self-selects" for something like parents' SES, which the person obviously did not choose. But the comparison is with the investigator, who clearly did not determine the level for that variable, so it is as if the person determined the level. At any rate, something about the person that the investigator had no control over was allowed to vary without the investigator's selection, or even knowledge, of the person's level on the variable.)

Any time a study merely measures, as opposed to manipulates, a given variable, we have to recognize that the subject and not the investigator has selected the level of the variable in question— along with the levels on all other variables measured or not. This gives up a huge degree of inferential power. In the class-size example, the investigator using multiple regression has allowed the level on the class-size variable to be self-selected (that is, the investigator did not determine class size), and the class-size variable may be associated with all sorts of other variables that may amplify or block the effects of class size on achievement. The only way to completely avoid the self-selection problem is for the investigator to select the value on the independent or predictor variable (for example, big class versus small class) and then observe its effects on the target variable (for example, achievement test performance). Alas, this is not always possible, so we have to be content with correlational analyses and multiple-regression analyses, hedged with caveats about the self-selection problem.

Finally, *statistical significance* tells us the likelihood that a result—for example, an effect of class size on performance—could have occurred by chance if the true effect is actually zero. The conventional value for statistical significance is .05, meaning that a difference between two means, or a correlation of a given size, would occur by chance only 5 in 100 times, or 1 in 20 times, in a study having the same design as the study in question. Statistical significance is very much a function of the number of observations. Even differences so small as to be of no practical or theoretical significance can be statistically significant if there are enough observations. Every result based on a study that I report in this book is statistically significant at least at the .05 level, except in one instance where I report a result that is "marginally significant," with a probability of less than .10 that the result would have been obtained by chance.

The Case for a Purely Environmental Basis for Black/White Differences in IQ

IN THIS APPENDIX I review and dispute the evidence that the IQ gap between blacks and whites is substantially genetic in origin. The case for genetics is presented in the chapter on race and intelligence in *The Bell Curve* by Richard Herrnstein and Charles Murray (1994) and in the recent review article by Phillipe Rushton and Arthur Jensen (2005). But many other scientists would also endorse at least some of the contentions below about genetic determination.

1. The heritability of IQ within the white population is high. Hence, it is probable that a very large fraction of the black/white gap is also genetic in origin.
2. The gap is not due to possible cultural differences because it is greater for allegedly culture-fair tests, such as those involving analysis of the relationship between geometric shapes, than it is for culture-loaded items, such as finding the analogy between *cotillion* and *square dance*.
3. Blacks from sub-Saharan Africa have even lower IQs (75 according to Herrnstein and Murray, 1994; 70 according to Rushton and Jensen, 2005) than African Americans. The reason

for the superiority of African Americans (with a putative IQ of 85) is that the gene pool of African Americans is about 20 percent European.

4. Blacks do relatively worse on items with high loadings on the so-called *g* factor (for general intelligence), and high *g*-loading items are more influenced by genetics than are low *g*-loading items.

5. Inbreeding depression (the detrimental results for IQ for the offspring of close relatives) affects performance on some items on IQ tests more than others. These items show the greatest difference between black and white scores, thus indicating that the most genetically influenced parts of the test show the biggest black/white differences.

6. Cranial capacity—brain size—is correlated with IQ within the white population and within the black population, and whites have greater cranial capacity than blacks.

7. Reaction time—in a setup in which people place their finger on a button and then move to touch a lighted bulb as quickly as possible—is lower for people with higher IQs and for whites than for blacks. (Low reaction time means high speed of reaction.)

8. Because black IQ is lower on average than white IQ for genetic reasons, the IQ in children of black parents with high IQs should regress to a lower mean than the IQ of children of white parents having the same IQs as the black parents. And in fact it does.

9. In a racial ancestry study examining black, white, and mixed-race children who were adopted by mostly middle- and upper-middle-class whites, the IQs of the black children differed little from those of the black population at large, and the IQs of the mixed-race children were in between those of the black children and the white children (Scarr and Weinberg, 1976, 1983).

Let's examine these contentions in order.

Heritability of IQ

The Bell Curve presents an argument that seems to settle the race and IQ debate in a simple and elegant manner. Those who think that blacks and whites are genetically identical, at least as far as intelligence goes, must believe that blacks can be treated as if they were simply a sample of whites selected for poor cognitive environment. Now if that is true, we want a measure of how strongly environment affects IQ, which is to say we want a correlation.

Studies on twins supply a correlation. Herrnstein and Murray assume a value that they believe to be a minimum probable influence of environment on IQ; that is, they assume that 60 percent of IQ variance is due to genetic differences between individuals, and 40 percent is due to environment. To get a correlation between environment and IQ, we have to take the square root of the percentage of IQ variance it explains. The square root of .40 is .63, so that is the correlation we want. Now we can calculate how far the environment of the average black would have to be below the environment of the average white to explain the IQ gap of 1 SD that separates the races: 1 SD divided by .63 gives us 1.59 SDs as the necessary environmental difference between the races. In sum, it takes that kind of environmental handicap to explain the racial IQ gap! How large this is can be appreciated by looking at a table of percentages under the mean of a normal curve. Once you reach 1.59 SDs below the mean, only 6 percent of the population is left—which is to say that the average black environment would have to be so bad that the environment for only 6 percent of American whites fell below it.

To Herrnstein and Murray, that was implausible, but worse was yet to come. Jensen (1998) gave a more up-to-date analysis of the twin studies. It shows that at adulthood, only 25 percent of IQ variance is due to environment. The square root of .25 is .50, and 1 SD (of IQ difference) divided by .50 is 2 SDs. So now we have to posit the average black environment as 2 SDs below the average white environment, which means that the environment for only 2.27 percent of American whites would fall below the average black environment. That is truly implausible.

The flaw in this argument is that it assumes that the environmental factors operating within groups are identical to those operating between groups. Geneticist Richard Lewontin illustrated this flaw. Imagine dividing a sack of seeds randomly into two groups, which means that while there will be plenty of genetic variation within each group, there will be no average genetic difference between the groups. You put all of the seeds of Group A into the same ideal potting mix, and you put all of the seeds of Group B into the same potting mix except for each plant in Group B you leave out one favorable ingredient. Clearly, within each group, height differences in plants at maturity are due to genetics—after all, there are zero environmental differences within the groups. But between the groups, the average difference in height is solely due to an environmental factor, namely, the fact that one group is missing a favorable ingredient in its potting mix.

So now we seem to have proved that twin studies within groups (you have no twin pairs of which one is black and the other white) have no implications for the potency of environmental factors between groups. But there is a problem. Can we imagine anything in the real world analogous to the missing ingredient in Group B's environment? It would have to be something disadvantageous that affected every member of Group B equally and not one person in Group A. Even the effects of racism could not do that. Some blacks would suffer from poverty and other disadvantages much less than others, and some whites certainly suffer from these things more than others. Lewontin's ingredient was given the name "factor X" to dramatize its mysterious character and imply that no one could think of anything that could play its role.

Dickens and Flynn (2001) proposed a formal model that solved the problem. They showed that two groups could be separated by an environmental factor of great potency that did not affect each member of the groups equally. They illustrated

the point with the effects that TV had on the escalation of basketball-playing ability. The new popularity of the game meant
that young people played much more and developed new technical skills, such as passing and shooting with both right and left
hands. When graduates came back to play their old high school
teams, they were soundly defeated even though they suffered no
genetic disadvantage (they were just as tall and quick). Obviously, the new basketball environment did not separate the two
groups neatly; some had practiced a lot even before TV. And
every member within each group was not affected to the same
degree. Within an age group, those with genes that made them
taller and quicker would be more likely to take to basketball and
get the advantage of team play and coaching than those who
were short and stout. Indeed, that is the reason why twin studies are deceptive. With their identical genes and identical height
and quickness, identical twins were likely to have common basketball histories—and since genes and powerful environmental
factors went together, genes would get the credit for the whole
package.

It is now easy to imagine a variety of environmental factors
that are potent and that separate black and white Americans.
Different child-rearing practices, different youth cultures, and so
forth could have a powerful effect on how much each group does
"mental exercise" and on the cognitive problem-solving skills they
each develop. And none of these things has to behave like the
implausible factor X.

In Chapter 6, I detailed the powerful environmental factors
that differ greatly between blacks and whites on average. It can
no longer be denied that such differences are capable in principle
of accounting for the race difference in IQ. The fact that the difference between blacks and whites now stands at two-thirds of a
standard deviation instead of at one standard deviation gives us
still further reason to believe that the differences are environmentally produced. Certainly the reduction in the difference has nothing to do with genetics.

Culture-Saturated and Culture-Fair IQ Test Items

On the surface, the observation that blacks score worse on alleg-edly culture-fair items than on culturally saturated items is a difficult one to counter. But recall from Chapter 3 (on changing IQ scores over time) the study by James Flynn (2000a) which showed that it is precisely on those IQ tests and subtests gener-ally deemed as "culture fair" that scores have increased most over the years. For example, the gain on the allegedly culture-fair Raven Progressive Matrices has been enormous, far outstripping the gain on more apparently culturally loaded tests. The gain on subtests such as vocabulary and information has been far less than that on allegedly culture-fair tests such as block design (which simply involves manipulating geometric patterns) and the other performance or fluid-intelligence subtests, including object assembly and picture arrangement. Since we know that the gains in IQ over a period as short as a generation could not possibly be due to genetic causes, we have to assume that something environmental is improving the performance on more allegedly culture-fair subtests than on more apparently culture-saturated subtests. Thus the culture-fair argument gets turned on its head. It is precisely the tests that we now know to be the most cultur-ally saturated, such as the Raven and block design, that most differentiate blacks and whites.

Sub-Saharan Africans Have IQs of 70 or 75

Let's stop and think for a moment about what an IQ of 70 might mean, if we took it seriously as an actual indicator of intelligence of sub-Saharan blacks. That figure is lower than the IQ for all but the lowest 2 percent of whites. Given what we know about people with such low IQs in our society, the average African, then, might not be expected to know when to plant seeds, what the func-tion of a chief might be, or how to calculate degrees of kinship. Obviously something is desperately wrong with these African IQ scores. They cannot possibly mean for the African population what they do for people of European culture.

The low scores rest primarily on data summarized by Richard Lynn and Tatu Vanhanen (2002), and are based largely on the (highly environmentally responsive) Raven Progressive Matrices. The samples used by Lynn and Vanhanen are generally small and haphazard. Moreover, they ignore samples with IQ means that are relatively high (Wicherts, Dolan, Carlson, and van der Maas, 2008). The test scores tell us little or nothing about the actual intelligence of Africans. Instead, they simply tell us that Africans have not yet undergone the gain in IQ scores, especially fluid-intelligence IQ such as that measured by the Raven matrices. Consistent with this, recall from Chapter 3 a recent study which showed that in a particular region of Kenya, Raven scores have gone up an unprecedented 1.75 SDs in a period of about fourteen years (Daley, Whaley, Sigman, Espinosa, and Neumann, 2003). Also note that a few months of Western-style education increased the scores of Africans on fluid-intelligence tests by .50 to .70 SD (McFie, 1961), and even a brief training session improved Raven scores in black Africans by an amount equal to 14 IQ points (while increasing the scores of whites by only 4 points) (Skuy et al., 2002).

Herrnstein and Murray (1994) and Rushton and Jensen (2005) acknowledge that the American "black" population contains about 20 percent European genes (Adams and Ward, 1973). In fact, they maintain, it is for this reason that the African American population has an IQ average of 85 rather than 70. Of course, following the simplest version of this logic, if the admixture of European genes in the black population were 40 percent instead of 20, then African Americans would have an average IQ of 100—equal to that of the white population—which requires 100 percent European genes. Continuing the argument to its absurd conclusion, if the admixture of European genes in the black population were 60 percent, then African Americans would have an average IQ of 115!

Blacks Do Worse on IQ Subtests
That Are Heavily g Loaded

When IQ test items (or subtests) are subjected to factor analysis, a tool for discovering the way in which the correlations in a matrix hang together, the first factor extracted is called *g* (for general intelligence). All items (or subtests) are correlated with this factor. A controversy exists as to whether *g* should be regarded as anything more than a statistical necessity for any set of items that are correlated among themselves to any degree. Some people treat *g* as being of little interest. Some invest it with great significance and point to its correlations with a number of physical and genetic variables, such as nerve conductance speed, as evidence that it is the principal engine of intelligence, with a substantial grounding in the physical nature of the nervous system. Herrnstein and Murray (1994) and Rushton and Jensen (2005) argue that because blacks and whites differ more in their performance on items and subtests that have higher *g* loadings (correlations with the *g* factor), this is evidence of the biological, genetic nature of the black/white difference in IQ.

The first thing to note about this argument is that it is based primarily on the loadings of the *g* factor on subtests of the Wechsler Intelligence Scale for Children (WISC). The differences in *g* loadings on the WISC are very slight, with the exception of a particularly low *g* loading for the Coding subtest. Otherwise the loadings are pretty much in the .60 to .70 range. It's scarcely impressive that there is a correlation between *g* loadings and black/white differences when the *g* loadings differ so little among themselves.

Flynn (2000a) pointed out that the WISC, on which the *g*-loading argument is largely based, is heavily tilted toward crystallized-intelligence subtests, including Information, Vocabulary, Comprehension, Arithmetic, and Similarities. If a test has a predominance of subtests of one kind, the first factor extracted will show heavy loadings for that kind of subtest. So if we were to use a lot of crystallized-intelligence subtests in our IQ test compendium, we would get out of our factor analysis a first, *g* factor that is heavily tilted toward crystallized intelligence.

But, says Flynn, suppose instead of looking at crystallized *g*, we were to look at fluid intelligence, or fluid *g*, which Jensen (1998) and other experts say is at least as much genetically influenced as crystallized *g*. Fluid *g* is measured by subtests such as those requiring the subject to create designs from geometric shapes or arrange pictures in a logical causal pattern. We can estimate the saturation of fluid *g* in the WISC subtests by determining their correlations with the Raven Progressive Matrices, a test that, according to Jensen and other experts in the field, is a virtually pure measure of fluid *g*. We now correlate the fluid-*g* ratings of each subtest with the sort of IQ gains that Flynn has reported as occurring in recent decades. What we find is that the higher the fluid-*g* loading for a subtest, the greater the gains in score over time on that subtest. We now have an absurdity: IQ gains, which are without question almost entirely environmental in origin, are found more on subtests that are allegedly more genetically influenced! Thus by knowing about the extent to which the gap is found for highly *g*-loaded subtests versus less *g*-loaded subtests, we learn nothing about the relative contribution of genes and environment to the black/white IQ gap. The gap is bigger for highly *g*-loaded subtests if we define *g* as crystallized, and this is supposed to show that the gap is genetic in origin. But if we define *g* as fluid, we find that the higher the *g* loading is, the more susceptible it is to environmental change. More genetically influenced subtests cannot be the ones that are the most environmentally influenced, so it is obvious that the argument from *g* loadings is flawed.

Finally, as I pointed out in Chapter 6, the contention that *g* loadings predict the magnitude of black/white differences on particular items entails the prediction that the scores on high *g*-loading items have changed the least for blacks over the past thirty years. William Dickens and James Flynn (2006) constructed a "*g*Q test"—an IQ score weighted by the *g* loading of each subtest on the WISC. Herrnstein and Murray and Rushton and Jensen would be obligated to say that even if blacks gained 5.5 points on IQ tests in general, they would not have gained very much on the heavily g-loaded tests. In fact, however, blacks have gained 5.13 points relative to whites on items weighted to reflect their g loadings defined as Jensen does.

It should be completely clear by now that the g-loading argument for genetically based IQ differences between the races is a red herring. The g loadings of subtests do not differ that much, the g loading of a particular subtest cannot be construed as evidence about the degree to which the subtest measures strictly biological or hereditary differences as opposed to environmentally produced differences, and scores for blacks have improved almost as much on a g-weighted IQ test as on a non-g-weighted test.

Blacks Do Worse on Subtests for Which Inbreeding Depression Is Relatively Great

The answer to this claim is the same in form as the answer to the claim about g loadings. On the face of it, if performance on a subtest particularly suffers from inbreeding depression, and if that subtest is particularly likely to differentiate between blacks and whites, this might seem like good evidence that the black/white gap is substantially influenced by biological, genetic factors. But it turns out that inbreeding depression for subtests, like fluid-g loadings, is also correlated with the extent of IQ gains over recent decades (Flynn, 2000a). In fact, the magnitude of the correlation is just as great for IQ gains as for black/white differences. Thus, we are confronted with another absurdity. If we are to believe that degree of inbreeding depression is an indicator of the genetic nature of the black/white differences in intelligence, we would also have to believe that degree of inbreeding depression indicates that IQ gains have a genetic cause. In other words, if we accept that degree of inbreeding depression is a measure showing the biological nature of the black/white difference, we also have to accept that degree of inbreeding depression shows that the Flynn effect—the increase in IQ over generations—is biological in origin.

Brain Size and the Black/White IQ Gap

The correlation between cranial capacity and IQ is probably about .30–.40 in the white population (McDaniel, 2005; Schoenemann, Budinger, Sarich, and Wang, 1999). Rushton and Jensen (2005) claim that cranial capacity for blacks is on average smaller than that for whites.

A difference between black and white brain size is not always found, however (National Aeronautics and Space Administration, 1978). More important, the correlation found within the white population probably does not indicate that greater brain size causes higher IQ. Within a given family, the sibling with the larger brain has no higher IQ on average than the sibling with the smaller brain (Schoenemann, Budinger, Sarich, and Wang, 1999).

In any case, as always, within-population differences do not necessarily tell us about the reasons for between-population differences. The fact that the smarter people within a given population have bigger cranial capacities does not tell us that the reasons for the size difference between blacks and whites are the same as the reasons for the size differences among people of the same race who differ in IQ. The male/female differences in cranial capacity are substantially larger than the black/white differences (Ankney, 1992). Yet the two genders have the same average IQ. (It should be noted that IQ tests are often engineered so that men and women come out with the same average score of 100. But the differences in average scores between the sexes on most test items are very slight so it is not difficult to engineer for gender equality.) Moreover, there exists a group of very short-stature individuals in Ecuador whose head size is several standard deviations below the mean (Guevara-Aguire et al., 1991). These individuals have not merely normal intelligence but unusually high intelligence (with a majority being among the highest ranking in their school class).

One large sample of blacks shows that the cranial capacity of black females was the same as that of whites, yet the IQ difference was the usual standard deviation typical of the gap at the time the data were collected (Joiner, in press). The IQ difference therefore is found in the absence of a cranial-capacity difference.

Finally, it is likely that the brain-size differences between blacks and whites that are sometimes found are environmental rather than genetic in origin (Ho, Roessmann, Hause, and Monroe, 1981). Pregnant black women are more likely to have any number of conditions that can result in smaller size of both body and brain, ranging from poor nutrition to alcohol use, than are white women. Perinatal factors are also much more negative for blacks in general than for whites (Bakalar, 2007); and prematurity is associated with much lower brain size, especially for black babies (Ho et al., 1981) It is only when babies are premature that the brains of black babies are smaller than those of white babies (Ho et al., 1981) Postnatal conditions also favor whites over blacks, especially for nutrition (Ho, Roessmann, Straumfjord, and Monroe, 1980).

So we do not learn much that is very compelling from the fact that blacks are sometimes found to have smaller brains than whites. Correlations within populations should not be extrapolated to between-population differences, and in any case, the within-population differences may not be due to greater brain size being causally related to higher IQ.

Reaction Times Are Slower for Blacks

More intelligent people in the white population have quicker reaction times than do less intelligent people. In addition, the variability of the reaction times for higher-IQ people is less, meaning that the reaction times of higher-IQ people are more uniform than those of less intelligent people. The correlations with IQ are low— around .20 (Deary, 2001)—and are not always found, but the best bet is that there are weak associations. And reaction times and variability of reaction times are longer and greater, respectively, for blacks than for whites (Rushton and Jensen, 2005).

Once again, let us note first of all that, as for brain size and any other variable associated with IQ within a population, the between-population differences do not necessarily have the same cause as the within-population correlations. Moreover, reaction time increases very little after the age of eleven, but intelligence keeps on growing apace (Nettelbeck, 1998). And some mentally retarded people have extremely fast reaction times (Flynn, 2007).

But these caveats are the least of the problems for the argument for black intellectual inferiority based on slow reaction times. First of all, both Herrnstein and Murray (1994) and Rushton and Jensen (2005) maintain that Asians have slightly higher IQs than whites, and both sets of authors imply that the reasons for that difference may be at least partially genetic. In Rushton and Jensen's Table 1, drawn from a book by Lynn and Vanhanen (2002), we learn that a sample of Hong Kong subjects has an average IQ of 113 and a sample of Japanese subjects has an average IQ of 110. (These estimates are much higher than what is commonly reported in the literature, incidentally, and recall from Chapter 8 that the evidence indicates East Asians do not have a higher IQ than Westerners.) The East Asians also have shorter and less variable reaction times than do the other groups of whites and blacks included in the table. However, Jensen (with Whang, 1993) reported that the reaction times and variabilities in a group of Chinese Americans were longer and greater, respectively, than those in a group of European Americans, even though the same Chinese Americans had an average IQ that was 5 points higher than that for the European Americans. And Lynn and Shighesia (1991) reported that although the reaction times in a group of Japanese were faster than those in a group of British subjects, the

Japanese had higher variabilities. Flynn (1991b) reported that it
was movement time and not reaction time that correlated with
IQ for the Chinese subjects. Movement time is a measure of how
long it takes a person to move a finger from the starting position
after he has made the decision to move it. Across a host of studies,
movement times are just as highly correlated with IQ as reaction
times (Deary, 2001). And blacks have faster movement times than
do whites! Rushton and Jensen (2005) do not mention any of
these complications to their simple reaction-time and race story.
In addition, the differences between the black South Africans and
the Irish reported by Lynn and Vanhanen are huge for IQ but
trivial for reaction times. In short, the overall results are a mess for
the set of contentions that (a) mean reaction time and variability
in reaction time are correlated with IQ, but (b) movement time is
not correlated with IQ, and (c) reaction times but not movement
times are faster for Asians than they are for whites, and (d) reaction times are faster for whites than they are for blacks but their
movement times are slower than those of blacks. The reasonable
position at this point is to assume that we know nothing of any
clarity or value about the interrelations among reaction time,
movement time, and race.

Black IQ Regresses to a Lower Mean than White IQ

Hereditarians often claim that because IQ is lower on average for
blacks than for whites for genetic reasons, the IQ of children of
black parents with high IQs should regress to a lower mean than
the IQ of the children of white parents having the same IQ as the
black parents. In other words, high IQs for blacks are farther from
the genotypic average of the black distribution than comparably
high IQs for whites are from the genotypic average of the white
distribution, so the IQs for children of high-IQ blacks have farther
to drop on average than do the IQs for children of high-IQ whites.
And it apparently is the case that the children of high-IQ blacks
have lower average IQs than do the children of comparably high-IQ

whites. This argument is quite weak because the same prediction can be derived from an environmental theory. If environmental factors such as parenting practices and subculture pressures toward low intellectual performances are pushing the average black IQ down more than the average white IQ, then we would also expect regression to a lower mean for the offspring of high-IQ blacks—for reasons having nothing to do with genetics.

Racial Ancestry and IQ

All of the research reported above is most consistent with the proposition that the genetic contribution to the black/white difference is nil, but the evidence is not terribly probative one way or the other because it is indirect. The only direct evidence on the question of genetics concerns the racial ancestry of a given individual. The genes in the U.S. "black" population are about 20 percent European (Parra et al., 1998; Parra, Kittles, and Shriver, 2004). Some blacks have completely African ancestry, many have at least some European ancestry, and some—about 10 percent—have mostly European ancestry. Does it make a difference how African versus European a black person is? A hereditarian model demands that blacks with more European genes have higher IQs. Herrnstein and Murray (1994) and Rushton and Jensen (2005), as it happens, scarcely deal with this direct evidence.

Children of different racial ancestry adopted into white families. Herrnstein and Murray (1994) and Rushton and Jensen (2005) reported on a study by Scarr and Weinberg (1983) showing that black children adopted by white families have a lower average IQ than white children adopted by white families, with mixed-race adoptees having an average IQ in between. Under the simplest model of pure genetic determination of the black/white IQ gap, the white adoptees should have had an average IQ 15 points or so higher than the average for black adoptees. The average for mixed-race adoptees should fall in the middle. When the children were about seven years old, their IQs were most consistent with a model of a very slight genetic contribution to the gap. When they were adolescents, their IQs suggested a larger genetic contribution (Weinberg, Scarr, and Waldman, 1992).

Scarr and Weinberg (1983) identified several factors that made their study a weak test of the genetic hypothesis. First, adoption agencies may have engaged in selective placement, which could have had the effect of putting the black adoptees into families that were of relatively lower social class. Second, since the natural parents' IQs were not known, it is possible that the natural parents of the white children had higher (genotypic) IQs than the white population in general, or the natural parents of the black children had lower genotypic IQs than the black population at large, which by itself could explain why the white adoptees had a higher average IQ than the black adoptees. Third, the black children were adopted at a substantially later age than the white children, and late adoption has negative implications for IQ. Fourth, the black children had more prior placements in foster homes, which is also associated with lower IQ. Fifth, the preadoptive placements of the black children were worse. Sixth, Sandra Scarr told me that the adolescent black and interracial children had an unusual degree of psychological disturbance having to do with identity issues. Some children reported, in effect, "I look in the mirror and I'm shocked to see a black person because I know I'm really white." Other children were disturbed because they felt that they were really black and didn't know why they had been consigned to an alien white family. As a consequence of all these problems, the authors cautioned against any conclusion with respect to the role of heredity in intelligence for adolescents. In any case, as we will now see, the Scarr and Weinberg study is the sole racial-mixture study that gives any support to the hypothesis that European genes make a "black" person smarter.

Black and white children raised in an enriched environment. Another study on black and white children raised in the same environment reached very different conclusions from those reached by Herrnstein and Murray (1994) and Rushton and Jensen (2005) on the basis of the Scarr-Weinberg study. This study was of black, white, and mixed-race children raised in an excellent institutional setting (Tizard, Cooperman, and Tizard, 1972). The caretakers were particularly well trained and conscientious, and the children's days were structured around highly intellectually stimulating activities. At four or five years old, white children had an average IQ of 103, black children had an average of 108, and children of mixed race had an average of 106. On their face, these results are most compatible with the assumption of a nontrivial genetic advantage for blacks. The black children in this study were West Indian and the white children were English. While it is possible that the black parents had unusually high genotypic IQs, Flynn (1980) argued that selective migration of West Indians to Britain could not have raised IQ scores by more than a very few points. Nevertheless, like the Scarr-Weinberg study, this one suffers from the fact that we do not know the IQs of the natural parents.

Black children adopted by black or by white families. Another adoption study had a design that was different from the Scarr-Weinberg study, and seems clearly superior to it. The study groups were black and mixed-race children raised in either black or white middle-class adoptive families (Moore, 1986). Black and biracial children raised by blacks had similar IQs, and so did black and biracial children raised by whites. Thus, having European genes was no advantage to the adopted children in either environment. Rushton and Jensen (2005) try to dismiss this study because the children were only seven years old at the time of testing. They say that "as people age, their genes exert ever more influence, whereas family socialization effects decrease (see Figure 3). Trait differences not apparent early in life begin to appear at puberty and are completely apparent by age 17" (p. 259). Their Figure 3, however, shows nearly identical heritability at ages seven and seventeen, so their own evidence refutes the idea that we can ignore Moore's finding of no difference between IQs in black and biracial children. More generally, there is plenty of evidence showing significant heritability of IQ by age seven, so the finding of no race difference at that age is quite telling.

The Moore study (1986) provides another test of environmental versus genetic hypotheses. Under the assumption that race differences in IQ are largely genetic, it should make little difference whether black or mixed-race children are raised by black or white families. Under the assumption that it makes a great deal of difference what kind of home, neighborhood, and school environment children grow up in, it should make a substantial difference whether the children were raised in black or white families. Thus, even though both the black and the white adoptive families were middle class, the investigator expected that the children raised in white families would have the higher IQs. And indeed this was the case. Children raised by blacks had an average IQ of 104, whereas those raised by whites had an average IQ of 117. Altogether the study provided no evidence consistent with a genetic contribution to the black/white gap in the population at large, and substantial reason to believe that the gap is largely or entirely environmental in nature. Once again, though, the missing key is the knowledge of the IQs of the natural parents. (And the number of subjects—forty-six—is lower than desirable.)

As it happens, there are additional studies that allow us to assess, even more directly than the adoption studies do, the question of the effects of European versus African ancestry. Any given member of the black population in the United States could have anywhere between 100 percent West African genes and mostly European genes. Are European genes good for the intelligence of blacks? Five qualitatively different kinds of studies provide answers to this question.

Studies of skin color. Studies relating darkness of skin color and IQ are easy to do, and many have been reported over the years. Let's pause a moment and think about what the correlation between skin color and IQ might be in the black population under the assumption of a purely environmental cause of the black/white IQ gap. We would expect lighter skin to be associated with a greater advantage for African Americans, resulting in higher SES and the educational and environmental advantages that go along with it. Thus, we would expect a positive correlation, perhaps as high as .20 or .30 or even more, between skin color and IQ. In fact, however, the literature consistently shows that the correlation of IQ with skin color in the black population is quite low. Even Audrey Shuey (1966), one of the most vehement supporters of the view that the black/white IQ gap is genetic in origin, reached the conclusion that IQ is only very weakly associated with skin color. Typical correlations are in the range of .10 to .15. Correlations between IQ and the degree to which facial features are rated as stereotypically African are similarly low (Shuey, 1966). Even if we ignore the advantages that might accrue to blacks with light skin, a correlation of .10 does not suggest that European ancestry exerts a strong genetic influence on IQ. On the other hand, many of the studies Shuey reviewed had small samples and dubious sampling procedures. Both skin color and IQ are measured with high reliability, but a major problem with these studies is that while skin color may seem to be a straightforward indicator of degree of European ancestry, it is not. Skin color varies substantially in sub-Saharan African populations. As a result, some Africans have relatively light skin for reasons that have nothing to do with European ancestry. A strong test of the "European ancestry" hypothesis therefore requires a more reliable indicator.

Studies measuring European ancestry via blood-group indicators. Fortunately, there are data available that reinforce the null implications of the skin-color studies. The frequency of different blood groups varies by race. Some blood groups that are common in European populations are rare in African populations and vice versa. Under the genetic hypothesis, blacks with more "European" blood groups should have more European genes and hence higher IQs. But Scarr, Pakstis, Katz, and Barker (1977) found that the correlation between IQ and degree of European heritage among blacks was only .05 in a sample of 144 black adolescent twin pairs. When skin color and socioeconomic status were controlled, the correlation dropped slightly, to –.02. It is important to note that these researchers found a typical correlation of .15 between skin color and IQ, suggesting that the comparable correlations in other studies were due not to slight superiority of European genes but to some other factor associated with light skin color in the black population such as social advantage.

Loehlin and colleagues (1973) correlated the estimated Europeanness of *blood groups* (rather than the Europeanness of individuals, estimated from their blood groups) with IQ in two different, small samples of blacks. They found a .01 correlation in one sample and a nonsignificant −.38 correlation in the other sample, with the more African blood groups being associated with the higher IQs.

It should be noted, however, that the blood-group studies are not as definitive as they might seem on the surface. This has to do with technical reasons related to the fact that white blood genes are only very weakly, if at all, associated with one another in the black population. If not associated with one another, then they might also not be associated with the white genes that are determinative of IQ.

Children born to black and white American soldiers in World War II. A German psychologist (Eyferth, 1961) studied the IQs of several hundred illegitimate children of German women fathered by black American GIs during the post-1945 occupation, and compared them to those fathered by white GIs. Again, take a moment to do a thought experiment here. We know there was very substantial prejudice against the mixed-race children, since it would have been obvious that they had been fathered illegitimately by foreign soldiers. We thus would expect, even under the hypothesis of zero genetic contribution to black/white IQ differences, that the mixed-race children would have suffered disadvantages that could have contributed to a lower IQ. But in fact the children fathered by black GIs had an average IQ of 96.5 and the children fathered by white GIs had an average IQ of 97. Inasmuch as the (phenotypic) black/white IQ gap in the military as a whole was close to that in the general population, these data imply that the black/white gap in the U.S. population as a whole is not genetic in origin (Flynn, 1980, pp. 87–88). These data also are not quite as probative as might appear on the surface, because the Army used a cutoff for IQ in accepting soldiers and that cutoff excluded a higher portion of blacks than whites, meaning that blacks were an unrepresentatively elite group. Flynn (1980) estimated that this could have produced no more than a 3-point difference in IQ between the black Army population genotype and the genetic composition of the black population as a whole, and probably less, but that loophole means that the study results are less than definitive. (It should be noted that some of the children were those of North African troops. Flynn (1980), however, estimated that this could affect expectations about the IQ of children born to soldiers of color by only a very small amount—unless one assumes that the average genotypic IQ for the North African soldiers was far higher than that known for any military group.)

Effect of white ancestry. A third approach to estimating blacks' white ancestry is to ask them about their family history. Imagine a 15-point black/white difference in IQ that is largely genetic in origin. Then think of four groups of blacks: one has only African ancestry, one has more African than white ancestry, one has equal African and white ancestry, and one has more white than African ancestry. Under the assumption of any contribution of genetics to the black/white IQ difference, those groups should differ in IQ. If we singled out blacks of particularly high IQ, we would expect to find that a very disproportionate number of them would have substantial white ancestry.

Witty and Jenkins (1934, 1936) identified from a sample of black Chicago schoolchildren sixty-three with IQs of 125 or above and twenty-eight with IQs of 140 or above. On the basis of their self-reports about ancestry, the investigators classified the children into the four categories of Europeanness just described. The children with IQs of 125 or above, as well as those with IQs of 140 or above, had slightly *less* European ancestry than the best estimate for the American black population as a whole at the time. This study was not ideal. It would have been better to compare the degree of European ancestry in high-IQ Chicago children to that of other black Chicago children rather than to the entire black population. But once again the results are consistent with a model of zero genetic contribution to the black/white gap or, perhaps, a slight genetic advantage for Africans.

Mixed-race children born to white versus black mothers. If the black/white IQ gap is largely genetic, children of mixed parentage should have the same average IQ regardless of which parent is black, since there is no a priori reason to assume that the genotypes of the black mother/white father children and the white mother/black father children would be different. (Though it should be noted that black fathers in the study had somewhat higher education and occupational levels than did black men generally, which could have meant that the genotypes for the children of white mothers and black fathers were slightly superior to those for the children of black mothers and white fathers.) But if (a) mothers are more important than fathers to the intellectual socialization of their children, and if the socialization practices of whites favor the acquisition of skills that result in high IQ scores, and/or (b) if the child's peers are more likely to be white if the mother is white, then the children of white mothers and black fathers should score higher than the children of black mothers and white fathers. In fact, a study by Willerman and colleagues (1974) found that children of white mothers and black fathers had a 9-point IQ advantage over those of black mothers and white fathers. This result suggests that most if not all of the black/white IQ gap is environmental (but note that the children tested were only four years old, an age at which IQ scores predict adult IQ scores only modestly).

So what do we have in the way of studies that examine the effects of racial ancestry—by far the most direct way to assess the contribution of genes versus the environment to the black/white IQ gap? We have one flawed adoption study with results consistent with the hypothesis that the gap is substantially genetic in origin, and we have two less-flawed adoption studies, one of which indicates slightly superior African genes and one of which suggests no genetic difference. We have dozens of studies looking at racial ancestry as indicated by skin color and "negroidness" of features that provide scant support for the genetic theory. In addition, three different studies of Europeanness of blood groups, using two different designs, indicate no support for the genetic theory. One study of illegitimate children in Germany demonstrates no superiority for children of white fathers as compared to children of black fathers. One study shows that exceptionally bright "black" children have no more European ancestry than the best-available estimate for the population as a whole. And one study indicates that it is more advantageous for a mixed-race child to be raised by a family having a white mother than by a family having a black mother.

All of these racial ancestry studies are subject to alternative interpretations. Most of these alternatives boil down to the possibility that there was self-selection for IQ in black-white unions. If whites who mated with blacks had much lower IQs than whites in general, their European genes would convey little IQ advantage. Similarly, if blacks who mated with whites had much higher IQs than blacks in general, their African genes would not have been a drawback. Yet the extent to which white genes contributing to mixed-race unions would have to be inferior to white genes in general, or black genes would have to be superior to black genes in general, would have to be very extreme to result in no IQ difference at all between children of purely African heritage and those of partially European origin. Moreover, self-selection by IQ was probably not very great during the slave era, when most black-white unions probably took place. It is unlikely, for example, that the white males who mated with black females had on average a lower IQ than other white males. Indeed, if such unions mostly involved white male slave-owners and black female slaves, which seems likely to be the case (Parra et al., 1998), and if economic status was slightly positively related to IQ (as it is now), these whites probably had IQs slightly above average. The black female partners were not likely chosen on the basis of IQ, as opposed to comeliness. Similarly, it scarcely seems likely that either black or white soldiers in World War II were selecting their German mates on the basis of IQ.

Several studies, moreover, are immune to the self-selection hypothesis. In particular, the study involving black and white children raised in an institutional setting, and the study involving black children adopted into either black or white middle-class homes, could not be explained by self-selection for IQ in mating.

In short, though one would never know it by reading Herrn-stein and Murray's book (1994) or Rushton and Jensen's article (2005), the great mass of evidence on racial ancestry—the only direct evidence we have—points toward no contribution at all of genetics to the black/white gap.

IQ Gains for Blacks?

Despite this strong evidence against the genetic hypothesis, we would be reluctant to abandon it totally if there were no evidence of IQ gains in the past generation or so. Things have improved both materially and socially for blacks, what with the civil rights movement, affirmative action, the increase in the proportion of blacks who are middle class, and the fact that blacks have pen-etrated into the highest levels of the society (including a chairman of the Joint Chiefs of Staff, two secretaries of state, a serious candidate for the presidency, CEO of the largest media company in the world, and CEO of one of the largest brokerage houses in the world). Have these changes been accompanied by increases in IQ for the average black? The answer offered by Rushton and Jensen in their 2005 article is that they have not. The difference, they maintain, has been steady for nearly one hundred years at 1.1 SDs, or approximately 16.5 points.

But William Dickens and James Flynn (2006) have shown that between 1972 and 2002, the IQ gap between American blacks and American non-Hispanic whites decreased by 4.5 to 7.0 points when they were tested before the age of twenty-five (the range being dependent on the type of test used). The value Dickens and Flynn prefer is 5.5 points. As I noted in Chapter 3, tests have to be restandardized from time to time because they become outdated. To the extent possible, testers try to obtain random samples of the population for their standardization studies. Dickens and Flynn analyzed data from nine standardizations of four of the most commonly used IQ tests, ones they believe provide the more valid estimates of temporal changes in the IQ gap: the WISC, the Wechsler Adult Intelligence Scale, the Stanford-Binet, and the Armed Forces Qualification Test. Data collected over the full thirty years for the WISC showed a reduction of 5.5 points in the gap. Other standardizations for the other tests were done over briefer periods, but projections showed that they reached the same conclusion on average. This amounts to a reduction of one-third of a standard deviation, or about one-third of the difference between blacks and whites, over a thirty-year period.

Dickens and Flynn rejected consideration of using five other IQ tests for which restandardizations had been completed because they believed there were flaws in the sampling or design of the studies. Rushton and Jensen (2006) believe that four of those tests should have been included, all of which showed lower estimates of gain than the tests that Dickens and Flynn examined. However, if we analyze all nine tests for which there exist data at two or more points in time, we obtain a median gain value of 4.5, not very different from the estimate of Dickens and Flynn.

If we put together the fact that the average IQ for the population as a whole is gaining at the rate of 9 points per thirty-year generation with the fact that blacks have gained about 5 points on whites over the past thirty years, we will see that the blacks of today have higher IQs than the whites of an earlier period in our history. Flynn (2008) has asked what average IQ contemporary blacks would score if they were to take the very first Wechsler Adult Intelligence Scale prepared in 1947–48 (and standardized on a white-only sample). That was about sixty years ago, a period during which the average IQ for the population as a whole increased by 18 points. He calculates that today's blacks would outscore the whites of 1947–48 by about 4 points.

Academic Achievement Gains

But the black/white gap in IQ is not the only one that has been substantially reduced over time. Extremely good data show changes in the black/white gap in reading and math abilities, which are good indicators of intellectual competence and which we care about as much as IQ. Every few years, the U.S. Department of Education gives a test called the National Assessment of Educational Progress to a random sample of children aged nine, thirteen, and seventeen. We can look at the gap in reading and math for children born as early as 1954 and as late as 1994. For the cohorts initially tested, reading scores for blacks were greatly lower than those for whites, with the gap ranging about 1.1 to -1.2 SDs on average. For the most recent cohorts, the gap is between .60 and .80 SD—a very large reduction. It should be noted that progress was not uniform across the time period. In the early years there was an astonishingly high rate of reducing the gap, followed by a nontrivial increase in the gap in the middle years. Only recently has the gap started decreasing again. I know of no convincing explanations for why the rate of improvement was so high for children born in the mid-1950s to the early 1970s, or for why the improvement reversed itself somewhat for the children born slightly after that time, or for why the improvement has begun again.

There is also good news on the black/white gap in math achievement, which measured a full 1.2 SDs for the first groups of children to take it. There was a dramatic improvement for children born between the mid-1950s and the late 1960s, a leveling-off or slight increase in the gap for children born between the early 1970s and the late 1980s, and renewed improvement for children born since that time. Again, no clear explanation exists for any of these changes in the trend line, but the overall situation is good: a reduction of more than a third of the gap to the point where the math gap is in the range of .60 to .90 SD.

It is interesting to note that if we convert the National Assessment of Education Progress gains to IQ-type scales, with the mean set to 100 and standard deviation to 15, and average the gains in math and reading for nine-, thirteen-, and seventeen-year-olds, we obtain an estimate of a 5.4-point reduction in the black/white gap during the period for which Dickens and Flynn (2006) found a reduction of 5.5 in the black/white gap for IQ.

In sum, the indirect arguments for genetic determination of the black/white gap in IQ are inherently weak and readily refuted. The most direct evidence—the only evidence that really counts—concerns the European heritage within the black population. With a single exception—which happens to be the only study reported on at any length by either Herrnstein and Murray (1994) or Rushton and Jensen (2005)—the data show that more European genes are not advantageous for blacks. The last thirty years have seen a reduction in the gap in IQ by about a third and a reduction in the academic achievement gap by about the same amount. The evidence favors a completely environmental explanation of the remaining difference between blacks and whites.

NOTES

Chapter 1: Varieties of Intelligence

1 "By intelligence the psychologist": Burt, Jones, Miller, and Moodie, 1934, p. 28.

4 "[Intelligence is] a very general mental": Gottfredson, 1997, p. 13.

4 Experts in the field of intelligence: Snyderman and Rothman, 1988.

4 Developmental psychologist Robert Sternberg has studied: Sternberg, 2007b.

4 In addition, East Asian understanding of intelligence: Nisbett, 2003.

7 These include "working memory,": Baddeley, 1986.

9 In addition, the subtests: M. J. Kane and Engle, 2002; Kazui, Kitagaki, and Mori, 2000; Prabhakaran, Rypma, and Gabrieli, 2001.

9 and another region linked: Rueda, Rothbart, McCandliss, Saccomanno, and Posner, 2005.

9 The destruction of the PFC has devastating consequences: Blair, 2006; Duncan, Burgess, and Emslie, 1995.

10 As one would expect given the lesion evidence: Blair, 2006; Prabhakaran, Rypma, and Gabrieli, 2001.

10 Additional evidence: Braver and Barch, 2002; Cavanaugh and Blanchard-Fields, 2006; Raz et al., 1997.

11 That fluid intelligence declines: Raz et al., 1997.

11 A final source of evidence: Jester et al., 2008.

11 Fluid intelligence is more important: Blair, 2006.

12 Over time, continued stress: Blair, 2006.

12 IQ tests tend to measure: Neisser, 1996; Sternberg, 1999, 2007a.

12 Robert Sternberg measures practical intelligence: Sternberg, 1999, 2006, 2007a.

13 Sternberg also writes about: Sternberg, 1999.

13 When Sternberg measures analytic intelligence: (Sternberg, 1999, 2006; Sternberg, Wagner, Williams, and Horvath, 1995.

14 Howard Gardner argued: Gardner, 1983/1993.

14 These include various "personal intelligences": Lopes, Grewal, Kadis, Gall, and Salovey, 2006.

14 Emotional intelligence as measured by Salovey: Lopes, Grewal, Kadis, Gall, and Salovey, 2006.

15 Decades ago, personality psychologist Walter Mischel: Mischel, 1974.

15 Mischel then waited more than a decade: Mischel, Shoda, and Peake, 1988.

16 Psychologists Angela Duckworth and Martin Seligman: Duckworth and Seligman, 2005.

17 To these qualifications of the importance of IQ: Rothstein, 2004.

18 Political scientist Charles Murray has looked at people: Murray, 2002.

20 Murray himself has long been associated: Herrnstein and Murray, 1994.

Chapter 2: Heritability and Mutability

21 "75 percent of the variance [in IQ]": Jensen, 1969, p. 1.

21 "Being raised in one family": Scarr, 1992, p. 3.

21 Some still consider it: Bouchard, 2004, Plomin and Petrill, 1997.

21 This environmentalist camp estimates heritability: Devlin, Daniels, and Roeder, 1997; Otto, 2001; Stoolmiller, 1999.

23 The researchers I call the strong hereditarians: For example, see Bouchard, 2004.

23 Have a look at Table 2.1: Devlin, Daniels, and Roeder, 1997.

24 This figure is .74, and is essentially the one that Arthur Jensen: Jensen, 1969. If you are sophisticated in your knowledge about correlations but not about hereditability studies, you may wonder why the correlation of .74 is not squared in order to derive an estimate of the percent of variance accounted for by genetics. The answer is that genetic correlations are already squared.

25 This correlation is .26: Bouchard and McGue, 2003. Another estimate of the effect of the environment comes from the difference between the correlation between parents and the children they raised (.41) and the correlation between parents and their children who were raised by someone else (.24). This difference is .17, which is pretty close to .20. And we can compare the correlation for siblings reared together and who therefore share the same environment (.46), with the correlation for siblings raised apart and who therefore do not share the same environment (.24). This comparison gives us an estimate of .22, which is also pretty close to .20.

25 This is because when people: McGue and Bouchard, 1998.

26 Billy is likely to be raised: Bronfenbrenner, 1986, 1975/1999.

26 Developmental psychologist Urie Bronfenbrenner: Bronfenbrenner, 1975/1999, 1986.

26 When environments are dissimilar: Taylor (1980) also found that similarity of the environment makes a great deal of difference to the correlation of IQ between identical twins, but Bouchard (1983), using IQ tests different from those Taylor used, reached the conclusion that similarity of the environment does not make much difference to the correlation of IQ between identical twins.

27 Devlin and his collegues: Devlin, Daniels, and Roeder, 1997.

27 Once corrections are made: Devlin and his colleagues (1997) also want to subtract 20 percent from the heritability estimates, owing to what they claim is the common environment in the womb for twins. This has proved to be controversial, however, and it remains to be seen how important the womb-similarity factor is in contributing to the similarity in intelligence between twins.

27 Developmental psychologists Sandra Scarr: Dickens and Flynn, 2001; Flynn, 2007; Scarr and McCartney, 1983.

29 Psychologist Mike Stoolmiller: Stoolmiller, 1999.

29 First, the socioeconomic status (SES): Maughan and Collishaw, 1998; Verhulst, Althaus, and Versluis-den Bieman, 1990.

30 Stoolmiller calculated that: Stoolmiller, 1999. Behavioral geneticist Matt McGue and his colleagues (2007) studied adoptive and nonadoptive families and found evidence for only a modest range restriction on socioeconomic status (SES) and psychopathology for adoptive families as compared to nonadoptive families. In addition, they found little evidence that adoptive sibling correlations for IQ were any higher when they corrected for such range restriction than when they did not correct for range restriction. However, these findings have to be interpreted in light of two facts. (1) All of their families, including the nonadoptive ones, had two adolescent children living at home, and such families are more stable and are of a somewhat higher SES than families in general. (2) Mothers in nonadoptive families that refused to participate had dramatically lower education levels than did the mothers of the participating nonadoptive families. Thus, the nonadoptive families who participated in the study were of higher SES and were probably more stable than nonadoptive families in general. As we will see later, heritabilities for such high-SES families are substantially higher than for the population at large.

30 Since we know that within-family variation: Stoolmiller (1999) showed, however, that the failure to estimate correctly the degree of variation within families has led to an overestimation of the contribution of within-family variation to IQ, just as it has led to an overestimation of the contribution of heredity to IQ.

31 Psychologist Eric Turkheimer and his colleagues: Turkheimer, Haley, Waldron, D'Onofrio, and Gottesman, 2003. In other work, Turkheimer and others found a similar modification of heritability by social class (Fischbein, 1980; Gray and Thompson, 2004; Harden,

Turkheimer, and Loehlin, 2006; Rowe, Jacobsen, and Van den Oord, 1999; Scarr-Salapatek, 1971). Other investigators failed to find this effect, however (Scarr, 1981).

Even when younger and older twins are sampled by the same method—typically by mailed questionnaire—and studied longitudinally, we would expect the subjects to be progressively harder to contact and to persuade to come to the testing site as they get older. This makes it likely that the older the subject is, the more likely the subject is to be of higher socioeconomic status—hence from a group with relatively high heritability.

31   We know from the work by Stoolmiller: Stoolmiller, 1999.

31   And in fact the environment: Turkheimer (personal communication) believes that the wide range of environments in lower-SES homes may be less important in determining low heritability than the fact that many such homes do not provide an environment sufficiently favorable for genes, and the differences among them from individual to individual, to express themselves.

31   This is because . . . there is a substantial bias: Dillman, 1978. This point applies to both major types of twin studies—those that estimate heritability based on the correlation between twins reared apart and those based on Falconer's formula comparing the correlation between identical twins reared together with the correlation between fraternal twins reared together—2(MZ r–DZ r). Note that the middle-class response bias is observed in research in Europe, including Scandinavia and Holland, where some of the most frequently cited studies of adult heritability have been done (Bergstrand, Vedin, Wilhelmsson, and Wilhelmsen, 1983; Dotinga, Schrijvers, Voorham, and Mackenbach, 2005; Jooste, Yach, Steenkamp, and Rossouw, 1990; Sonne-Holm, Sorensen, Jensen, and Schnohr, 1989; Van Loon, Tijhuis, Picavet, surtees, and Ormel, 2003).

32   Psychologists Christiane Capron and Michel Duyme: Capron and Duyme, 1989.

33   So the study showed: Jensen (1997) attempted to minimize the implications of this very important study by saying that the correlation between *g* loadings of WISC subtests and the magnitude of the difference between biological parents and offspring are higher than the correlation between *g* loadings and the magnitude of the difference between adoptive parents and offspring. In other words, subtests that measure the genuine article, namely, *g*, the most show a relation for biological parents and offspring, and the less *g*-revealing subtests show a relation between adoptive parents and offspring. Many things can be said about this. (1) The difference in factor loadings on *g* across WISC subtests is relatively slight, so the reanalysis does not much affect conclusions about the impact of adoption on intelligence. (2) More importantly, there is no difference at all in average *g* loadings between the subtests that show a big difference between high- and low-SES adoptive parents and their offspring and those

that show little difference. Both loadings are .71 on average. (3) The fact that the subtests that most differentiate between biological parents and their offspring have higher g loadings than the subtests that differentiate least is almost entirely due to the fact that the Coding subtest, which has by far the lowest g loading, is one of the tests that does not differentiate much between high- and low-SES biological parents and their offspring. If Coding is left out, the g loadings for the former average .79 and for the latter average .69—not much of a difference. (4) The WISC is a heavily verbally oriented test—that is, it measures mostly crystallized intelligence, and the g loadings differ primarily to the extent that they measure verbal ability as opposed to performance or fluid skills. (5) Jensen himself (1998) said that the purest measure of fluid g is the Raven Progressive Matrices test. If we examine fluid g as defined by subtest correlations with the Raven scores, we find that the correlation for biological parents and their offspring flips direction. It is now the tests with the highest g loadings that differentiate *least* between high- and low-SES biological parents and their siblings. So the question as to whether the genetic effect is more reflective of differences in g than the environmental effect is entirely a matter of deciding which is the real g—fluid or crystallized. See Flynn (2000a) for a more detailed explanation of these points in the context of race differences in IQ. More important than any of these points is that the adopted children of upper-middle-class parents did far better in terms of academic achievement than did the adopted children of lower-class parents. The results of their school achievement tests were substantially better and they were far less likely to be put back a grade. And Jensen has in other contexts expressed the view that academic achievement is highly reflective of g. In any case, we care more about school achievement than about IQ.

33 Another French study: Schiff, Duyme, Stewart, Tomkiewicz, and Feingold, 1978.

34 In another extremely important natural experiment: Duyme, Dumaret, and Tomkiewicz, 1999.

34 Stoolmiller showed: Stoolmiller, 1999.

35 A review that examined all: van IJzendoorn, Juffer, and Klein Poelhuis, 2005.

35 This estimate was derived by comparing: Hereditarians may complain that the IJzendoorn estimates (van IJzendoorn, Juffer, and Klein Poelhuis, 2005) of the effects of the adoptive-family environment are too high because they are based on young children and heredity exerts greater effects on older than on younger people—presumably because people are more able to choose their environments as they get older, and people with genes for higher IQ choose environments that will make them smarter. But IJzendoorn and colleagues (2005) found that age of testing—twelve years or younger versus thirteen to eighteen years old—made no difference to the estimate of the effects of adoption. This is the usual finding of the relationship between age and

heritability. Heritability is constant from childhood to late adolescence (McGue, Bouchard, Iacono, and Lykken, 1993).

35 As it happens, the difference: Capron and Duyme, 1989.

35 The crucial implication of these findings: It is not just the IQ of children born to lower-SES parents that is highly modifiable by the environment. The IQ of children born to upper-middle-class parents is modifiable too. Such children have a much lower average IQ when they are raised by lower-SES parents, 12 points lower to be exact (Capron and Duyme, 1989).

35 The environments of adoptive families: Stoolmiller, 1999.

36 So the relatively low correlation: As noted in the text of Chapter 2, some people claim that the very low correlations for IQ found in adulthood between people adopted into the same family establish that such between-family environmental differences as may exist are gone by adulthood, when people have the ability to choose their own environments and their genetic potential can fully exert itself. Genetically smart people, the argument goes, seek out smarter environments, and less genetically smart people drift into less smart environments, so the childhood environment is no longer very relevant to contemporary IQ. However, the only sort of correlational study that would be reliable for this conclusion is a longitudinal one, where the correlation between adoptive siblings is observed from childhood through adulthood—for the same sample of individuals. This would remove the problem of researchers typically looking at children from higher-SES adoptive famiies when they look at adult samples than when they look at childhood samples. In fact, there appear to be only two longitudinal studies, and though they both show a reduction in the magnitude of correlations between childhood and adulthood, the samples are small and the differences between child correlations and adult correlations are not significant (Stoolmiller, 1999).

36 Finally, since Herrnstein and Murray: Herrnstein and Murray, 1994.

36 Locurto reported: Locurto, 1990.

36 Judith Rich Harris, the author: Harris, 1998.

36 In his brilliant book: Pinker, 2002.

36 "Studies have shown": Levitt and Dubner, 2006, p. 157.

37 One study looked at the IQs: Scarr and Weinberg, 1976.

37 Similarly, the cross-fostering study: Capron and Duyme, 1989.

Chapter 3: Getting Smarter

39 "even a perfect education system": Murray, 2007a.

39 "a person's total score": Raven, Court, and Raven, 1975, p. 1.

39 *The Bell Curve*: Herrnstein and Murray, 1994.

40 Developmental psychologists Stephen Ceci and Wendy Williams: Ceci, 1991; Ceci and Williams, 1997.

40 Kids are deprived of school: Ceci, 1991; Jencks et al., 1972.
40 Much, if not most: Cooper, Charlton, Lindsay, and Greathouse, 1996; Hayes and Grether, 1983.
40 The very oldest study: F. S. Freeman, 1934.
40 Another early natural experiment: Sherman and Key, 1932.
41 School was delayed for Dutch children: DeGroot, 1948.
41 The IQs of such children: Ramphal, 1962.
41 The IQs of the children: R. L. Green, Hoffman, Morse, Hayes, and Morgan, 1964.
41 Two different groups of Swedish psychologists: Härnqvist, 1968; Husén, 1951.
42 In fact, studies in Germany and Israel: Baltes & Reinert, 1969; Cahan and Cohen, 1989.
42 Western-style education can have big effects: Ceci, 1991.
43 As little as three months: Ceci, 1991.
43 In America in 1900: Folger and Nam, 1967.
44 James R. Flynn has documented: Flynn, 1987, 1994, 1998.
44 In what follows I stick closely: Flynn, 2007.
44 Scores for eighteen-year-olds: Loehlin, Lindzey, and Spuhler, 1975.
44 And in any case, if completely test-naive people: Flynn, 2007.
45 For every child: Rutter, 2000.
45 The even gains across the distribution: Murray, 2007a.
47 The graph reveals: Flynn, 2007.
48 Developmental psychologist Clancy Blair: Blair, Gamson, Thorne, and Baker, 2005.
48 Developmental psychologist Wendy Williams: Williams, 1998.
49 Blair and his coworkers: See Blair and Razza, 2007.
49 But Figure 3.2 shows: Flynn, 2007.
49 Researchers have shown: C. S. Green and Bavelier, 2003.
49 Neuroscientists have shown: Jaeggi, Perrig, Jonides, and Buschkuehl, in press.
50 Researcher Rosario Rueda and her colleagues: Rueda, Rothbart, McCandliss, Saccomanno, and Posner, 2005.
50 As we might expect: Klingberg, Keonig, and Bilbe, 2002; Rueda, Rothbart, McCandliss, Saccomanno, and Posner, 2005.
50 The ADHD researchers: Klingberg, Keonig, and Bilbe, 2002.
51 Cognitive neuroscientist Adele Diamond: Diamond, Barnett, Thomas, and Munro, 2007.
51 Scores on two tests: Flynn, 2007.
53 Children can learn a lot: Johnson, 2005.
53 This seems understandable: National Endowment for the Arts, 2007.
53 On the other hand, we do know from other evidence: National Center for Educational Statistics, 2008. Tables and Figures: http://nces.ed.gov/quicktables/(retrieved August 14, 2008).
53 It is worth noting: Williams, 1998.

54 At the turn of the twentieth century: Blair, Gamson, Thorne, and Baker, 2005.
54 By 1983, more than: Williams, 1998.
55 It does seem to have come to a halt: Schneider, 2006.
55 In a particular region of Kenya: Daley, Whaley, Sigman, Espinosa, and Neumann, 2003.
55 A study on the Caribbean island of Dominica: Meisenberg, Lawless, Lambert, and Newton, 2005.
55 This seems inevitable: Johnson, 2005.
56 The gains in IQ make it clear: Lynn and Vanhanen, 2002; Rushton and Jensen, 2005.
56 Finally, the evidence speaks: Murray, 2007a.
56 He has also said: Herrnstein and Murray, 1994.

Chapter 4: *Improving the Schools*

57 U.S. students who score in the 95th percentile: U.S. Department of Education, 1998.
58 Such additional funds may be spent: Hess, 2006.
58 The classic case of this: Evers and Clopton, 2006.
58 Money by itself: Evers and Clopton, 2006.
58 Other evidence about the effect: Hanushek, 2002.
59 Such studies found as much as a one-third reduction: Howell, Wolf, Peterson, and Campbell, 2001.
59 When studies on the effects: Krueger, 2001; Krueger and Zhu, 2004; Ladd, 2002; C. E. Rouse, 1998.
60 Reasonable experiments: Bifulco and Ladd, 2006.
60 Unfortunately, charter schools: Hoxby and Murarka, 2007.
60 There is, however, some evidence: Hoxby, 2004; Hoxby and Rockoff, 2004.
61 The multiple-regression analysts: Hanushek, 2002.
61 On the other hand, economist Alan Krueger: Krueger and Zhu, 2004.
61 The children in the smaller classes: Krueger, 1999.
61 The effects persisted: Nye, Jayne Zaharias, Fulton, Achilles, and Hooper, 1994.
61 Indeed so, but the fact: Hanushek, Kain, O'Brien, and Rivkin, 2005; T. Kane, 2007.
61 Nor, surprisingly, is possession: Hanushek, Kain, O'Brien, and Rivkin, 2005.
61 The average difference in reading: Rockoff, 2004.
62 But note that most of the difference: Hanushek, Kain, O'Brien, and Rivkin, 2005; Jacob and Lefgren, 2005.
62 Defined in this way, 1 SD: Hanushek, Kain, O'Brien, and Rivkin, 2005; T. Kane, 2007; Rockoff, 2004.

62 Economist Eric Hanushek's estimate of the impact: Hanushek, Kain, O'Brien, and Rivkin, 2005.

62 A study on the effects: Pedersen, Faucher, and Eaton, 1978.

63 Education researchers Bridget Hamre and Robert Pianta: Hamre and Pianta, 2001.

65 Hamre and Pianta found, in an earlier study: Hamre and Pianta, 2001.

65 Principals know about the quality: Armor, 1976; Murnane, 1975.

65 But there is little evidence: Hanushek, Kain, O'Brien, and Rivkin, 2005.

65 Researchers who are aware: De Sander, 2000.

65 A common complaint: Rosenholtz, 1985.

65 Israeli researchers conducted: Lavy, 2002.

66 Teachers in . . . and they monitor student performance: Connell, 1996.

66 The usual claim is that schools: Muijs, Harris, Chapman, Stoll, and Russ, 2004.

67 On the other hand: Muijs, Harris, Chapman, Stoll, and Russ, 2004.

67 Despite the hundreds of millions: Mosteller and Boruch, 2002.

67 Research is mostly anecdotal: Cook, 2003.

68 These studies generally yield: Borman, Hewes, Overman, and Brown, 2003.

69 Educational psychologist Geoffrey Borman: Borman, Hewes, Overman, and Brown, 2003.

70 A particularly well-designed: Borman et al., 2007.

70 The third-party comparison studies: See also particularly rigorous tests by Thomas Cook and colleagues (1999, 2000).

71 These computer software systems: Kulik, 2003.

72 Education researcher: Robert Slavin (Slavin, 1995).

72 Slavin reported studies: Slavin, 1995.

72 There are a variety of ways: Slavin, 1995.

73 In an extremely welcome development: U.S. Department of Education, 2008.

74 Herrnstein and his coworkers: Herrnstein, Nickerson, Sanchez, and Swets, 1986.

75 In fact, Mark Lepper and his colleagues found: Lepper, Drake, and O'Donnell-Johnson, 1997; Lepper, Wolverton, Mumme, and Gurtner, 1993; Lepper and Wolverton, 2001.

Chapter 5: *Social Class and Cognitive Culture*

78 "the class structure of modern": H. J. Eysenck, 1973, p. 19.

78 The average IQ: Flynn, 2000b.

80 Although the available evidence: Pollitt, Gorman, Engle, Martorell, and Rivera, 1993.

246   *Notes*

80 It is not clear that nutrition differences: Rothstein, 2004.
80 Even if hunger is rare: General Accounting Office, 1999.
80 And there is evidence: Schoenthaler, Amos, Eysenck, Peritz, and Yudkin, 1991.
80 The effects of lead: Baghurst, 1992.
80 Children whose mothers drank: Centers for Disease Control and Prevention, 2007.
80 Lower-SES women: Streissguth, Barr, Sampson, Darby, and Martin, 1989.
81 Low birth weight: Hack, Klein, and Taylor, 1995.
81 Lower-SES mothers: Anderson, Johnstone, and Remley, 1999; U.S. Department of Health and Human Services, 2006.
81 For children with the most common: Anderson, Johnstone, and Remley, 1999; Caspi, 2007; Kramer, 2008; Luca, Morley, Cole, Lister, and Leeson-Payne, 1992.
81 One study finds: Der, Batty, and Deary, 2006. On the other hand, in one study (Kramer, 2008), mothers were encouraged to breast-feed exclusively, and this experimental study got the same results as the correlational studies.
81 Lower-SES people: Mills and Bhandari, 2003.
82 One such harmful circumstance: Rothstein, 2004.
82 Lower-SES children are more likely: Rothstein, 2004.
82 Compared with higher-SES parents: Dodge, Pettit, and Bates, 1994.
83 Developmental psychologist Vonnie McLoyd: V. McLoyd, 1998.
83 Early emotional trauma: Blair, 2006.
83 Income inequality in the United States: Economic and literacy statistics in this paragraph and the next come from a recent book on the American labor market by Richard Freeman (2007).
84 In contrast, the after-tax: Rothstein, 2004.
84 Reflecting the differences: Organisation for Economic Co-operation and Development, 2001.
85 The difference in reading and math skills: Micklewright and Schnepf, 2004.
85 In fact, the achievement gap: Ceci, 2007.
85 While their children: Lareau, 2003
86 Psychologists Betty Hart and Todd Risley: Hart and Risley, 1995.
86 Degree of encouragement: Brooks-Gunn & Markman, 2005.
87 Much of what we know: Heath, 1982, 1983.
87 Heath's study was conducted . . . but more recent studies: Lareau, 2003; Mikulecky, 1996.
87 In what follows . . . recent work of Annette Lareau: Laueau, 2003.
89 A Philadelphia study: Neuman and Celano, 2001.
90 The IQs and skills of middle-income children: Cooper, Nye, Charlton, Lindsay, and Greathouse, 1996.

90 One study found: Burkham, Ready, Lee, and LoGerfo, 2004.
92 In fact, there is some evidence: C. Rouse, Brooks-Gunn, and McLanahan, 2005.

Chapter 6: IQ in Black and White

93 "The taboo against discussing race": Sowell, 1994, p. 168.
93 "[Black] kids seem to": Ogbu, 2003, p. 78.
93 A millennium earlier southern Europeans: Sowell, 1994, p. 156.
93 though Julius Caesar: Churchill, 1974, p. 2.
95 In this chapter and in Appendix B: Rushton and Jensen, 2005.
95 There is plenty of evidence: Steele, 1997; Steele and Aronson, 1995.
95 When the test is presented: Steele, Spencer, and Aronson, 2002.
95 At least as late as 1980: Jensen, 1980.
96 The correlation between brain size and IQ may be: Schoenemann, Budinger, Sarich, and Wang, 1999.
96 And according to a number of studies: These are reviewed in Rushton and Jensen, (2005).
96 In fact, however, there is no such correlation: Shoenemann, Budinger, Sarich, and Wang, 1999.
96 Moreover, the brain-size difference: Ankney, 1992.
96 And a group of people in a community in Ecuador: Guevara-Aguire et al., 1991; Kranzler, Rosenbloom, Martinez, and Guevara-Aguire, 1998.
96 The direction of recent evolution: Beals, Smith, and Dodd, 1984; Brown, 1992; Brown and Maeda, 2004; Henneberg, 1988; Henneberg and Steyn, 1993, 1995; Schwidetsky, 1977.
97 About 20 percent of the genes: Parra et al., 1998; Parra, Kittles, and Shriver, 2004.
97 It turns out that light skin color: Shuey, 1966.
97 Tested in later childhood: Eyferth, 1961.
97 But when a group of investigators: Witty and Jenkins, 1934.
97 The blood group assays: Scarr, Pakstis, Katz, and Barker, 1977.
97 Similarly, the blood groups: Loehlin, Vandenberg, and Osborne, 1973.
98 The hereditarians cite a study: Scarr and Weinberg, 1983; Weinberg, Scarr, and Waldman, 1992.
98 A superior adoption study: Moore, 1986.
98 Psychologists Joseph Fagan and Cynthia Holland: Fagan and Holland, 2002, 2007.
99 Indeed, black IQ now: Dickens and Flynn, 2006.
99 In fact, we know: Dickens and Flynn, 2006.
100 The shrinkage of the gap: Dickens and Flynn, 2006.
101 Black family income: Rothstein, 2004.

101 The unwed mother rate: Camarota, 2007.
101 Affirmative action has likely played: Thernstrom and Thernstrom, 1997.
102 Employers believe that young: Moss and Tilly, 2001.
102 When black and white job applicants: Darity and Mason, 1998; Darley and Berscheid, 1967.
102 The white applicants: Pager, 2003.
102 Already in 1965: Moynihan, 1965.
102 In 2005, for blacks age twenty-five to twenty-nine years old, the ratio of females to males: U.S. Census Bureau, 2006.
103 Since we know that more: Flynn, 1980.
103 As a consequence at least: U. S. Office of Personnel Management, 2006.
103 Race may be: Myrdahl, 1944.
103 African anthropologist John Ogbu: Ogbu, 1978, 1994.
103 The IQ differences between: Sowell, 1994.
104 These groups include whites: Ceci, 1991.
104 Ogbu focuses on: Ogbu, 1991a.
104 Unlike lower-caste minorities: Sampson, Morenoff, and Raudenbush, 2005.
104 Ogbu holds that: Ogbu, 1978.
104 In the case of American blacks: Ogbu, 2003.
104 Ogbu has written: For example, see Ogbu, 1991b.
105 In what follows I draw: Sowell, 1978, 1981, 1994; Flynn, 1991a.
105 In eighteenth-century Virginia: Sobel, 1987.
105 Though most free blacks: Sowell, 1978.
106 Eighty-five percent of free: Sowell, 1978.
106 In Chicago in 1910: Sowell, 1978.
106 To give an idea: Sowell, 1978.
108 The Irish, who were white: Ignatiev, 1995.
108 As of the mid-twentieth century: Macnamara, 1966.
108 English psychologist H. J. Eysenck: Eysenck, 1971.
108 The gene pool . . . and literary proficiency: Organisation for Economic Co-operation and Development, 2000.
108 Post–secondary school enrollment: Organisation for Economic Co-operation and Development, 2000.
109 The proportion of blacks: Attewell, Domina, Lavin, and Levey, 2004.
110 In 1970, second-generation West Indians: Sowell, 1978.
110 They took whatever jobs: Sowell, 1978; Waters, 1999
110 And West Indian culture: Sowell, 1978.
110 Sowell recently argued: Sowell, 2005. Sowell is worth quoting at length on this point. From page 6 of his book *Black Rednecks and White Liberals*: "The cultural values and social patterns prevalent among Southern whites included an aversion to work, proneness to violence,

neglect of education, sexual promiscuity, improvidence, drunkenness, lack of entrepreneurship, reckless searches for excitement, lively music and dance, and a style of religious oratory marked by strident rhetoric, unbridled emotions, and flamboyant imagery." From pages 1–2: "That culture long ago died out where it originated in Britain, while surviving in the American South. Then it largely died out among both white and black Southerners, while still surviving today in the poorest and worst of urban black ghettos."

111   I pointed out: Hart and Risley, 1995.
112   Recall from the last chapter: Heath, 1982, 1983.
113   This probably pays off: Loehlin, Lindzey, and Spuhler, 1975.
113   In fact, those abilities: Heath, 1982.
114   In the late 1980s: Heath, 1990.
115   Meredith Phillips, Jeanne Brooks-Gunn: Phillips, Brooks-Gunn, Duncan, Klebanov, and Crane, 1998.
115   One came from a study: Chase-Lansdale et al., 1991.
115   The second data set came: Brooks-Gunn et al., 1994.
115   Things studied include: M. Phillips, Brooks-Gunn, Duncan, Klebanov, and Crane, 1998, pp. 126–127.
116   Hart and Risley, in their study: Hart and Risley, 1995.
116   The three-year-old black child: Tough, 2007.
116   Recall the study: Moore, 1986.
118   These subcultures encourage: Patterson, 2006.

Chapter 7: Mind the Gap

119   "Compensatory education": Jensen, 1969, p. 1.
119   "There is no evidence": Jencks et al., 1972, p. 8.
119   "There is no reason": Murray, 2007a.
121   It results in mortality rates: Ludwig and Miller, 2005.
121   In earlier days, Head Start: McKey, Condelli, Ganson, McConkey, and Plantz, 1985.
121   and more recent studies: Grissmer, Flanagan, and Williamson, 1998; Ludwig and Miller, 2005.
121   Recent reports show lower effect sizes: U.S. Department of Health and Human Services, 2005.
121   What little there is: Ludwig and Miller, 2005. One study found that the benefits of greater high school completion and college attendance are limited to whites: Garces, Thomas, & Currie, 2002.
122   The cost of Head Start: For a pessimistic review of the literature on Head Start, see Besharov, 2005. For a more optimistic review of Head Start, see Ludwig and Phillips, 2007. However, the latter relies on study designs that I consider inferior. Another optimistic study is by Deming, 2008.

122   Early Head Start: Love, 2005.

122   A review of about: Grissmer, Flanagan, and Williamson, 1998.

123   The Perry Preschool Program: Schweinhart et al., 2005; Schweinhart and Weikart, 1980, 1993.

124   These advantages include: Barnett, 1992.

124   The fact that gains: Knudsen, Heckman, Cameron, and Shonkoff, 2006.

124   An intervention even more ambitious: Garber, 1988. Since the first edition of this book was published, it has been brought to my attention that the principal investigator of the Milwaukee Project was convicted of the crime of abuse of federal funding for private gain. The author of the book describing the project, which I summarize here, has never been accused of wrongdoing. However, the details of the project have apparently not been reported in a refereed journal. But the conclusions about the Milwaukee Project are entirely consistent with the conclusions of the Perry and Abecedarian programs.

126   A yet more intensive intervention: Campbell et al., 2001; Campbell and Ramey, 1995; C. T. Ramey et al., 2000; S. L. Ramey and Ramey, 1999.

129   One is important because: Herrnstein ad Murray, 1994.

129   Project Care, using methods: Wasik, Ramey, Bryant, and Sparling, 1990.

129   Another replication of Abecedarian: Gross, Spiker, and Haynes, 1997; Hill, Brooks-Gunn, and Waldfogel, 2003; The Infant Health and Development Program, 1990.

130   A particularly important fact: Gormley, Gayer, Phillips, and Dawson, 2005; Love et al., 2005.

130   and benefit poor children: Hamre and Pianta, 2005.

130   Some reasonably ambitious parenting interventions: Brooks-Gunn and Markman, 2005; Juffer, Hoksbergen, Riksen-Walraven, and Kohnstamm, 1997; van Zeigl, Mesman, van IJzendoorn, Bakersman-Kranenburg, and Juffer, 2006; Watanabe, 1998.

130   This was conducted: Landry, Smith, and Swank, 2006; Landry, Smith, Swank, and Guttentag, 2007.

132   The Heritage Foundation: Carter, 2000.

132   Richard Rothstein: Rothstein, 2004.

132   The Education Trust: Jerald, 2001.

133   Rothstein gives an even: Reeves, 2000.

133   The *New York Times* announced: Finder, 2005.

133   and yet again: Bazelon, 2008.

134   The black/white gap was actually: You can see the comparison of Wake County and statewide scores at the state's education Web site: http://www.ncreportcards.org/src/distDetails.jsp?Page=2&pSchCode=304&pLEACode=920&pYear=2004-2005&pDataType=1

134   By the new standards: Data for 2005/06 math scores are more interpretable because the cutoffs for proficiency scores are much lower than for 2004/05. As in many states, what is labeled "proficient" in North

Carolina bounces around from year to year depending more on politics than on actual student achievement. The 2005/06 data, when the required level for proficiency went up again, tell the same story. The Wake County black/white gap was no smaller than it was for the state as a whole. The *Times* writer would have done well to get his figures at the source rather than depending on data and interpretations offered by educators relying on the Wake County district reports.

134 There is good evidence: Murnane, Willett, Bub, and McCartney, 2006.

134 Experience in teaching counts, though: Sanders and Horn, 1996.

135 Again, there is the possibility: Sanders and Horn, 1996

135 And we know that: Hamre and Pianta, 2005.

135 We also know that: Grissmer, Flanagan, and Williamson, 1998.

135 The math training program: H. Phillips and Ebrahimi, 1993.

135 One study of Project SEED's effectiveness: Webster and Chadbourn, 1992.

136 Reading Recovery is a tutoring program: Slavin, 1995.

136 The Ohio State group conducted randomized studies: Slavin, 2005.

136 One independent study evaluating: Slavin, 2005.

136 There is at least one extremely: All of the information about KIPP comes from a report by the Stanford Research Institute: David et al., 2006.

137 KIPP maintains that "while the average fifth-grader": KIPP Web site, quoted in Mathews 2006.

137 However, SRI International conducted: David et al., 2006.

138 Principals believe: Quotations are from David et al., 2006.

141 The next step will be: Two other systems are similar to KIPP in design. One is called Achievement First and operates primarily in New Haven, Connecticut, and the other is called North Star and operates in Newark, New Jersey. These programs claim good results with children resembling the KIPP students. But they have not been well evaluated to my knowledge.

142 There's good news: Jessness, 2002.

143 They showed, not surprisingly: Blackwell, Trzesniewski, and Dweck, 2007; Henderson & Dweck, 1990.

143 Dweck and her colleagues: Blackwell, Trzesniewski, and Dweck, 2007.

143 Dweck reported that some: Personal communication, January 2007

143 Joshua Aronson and his colleagues: Aronson, Fried, and Good, 2002.

143 One study was conducted: Good, Aronson, and Inzlicht, 2003.

144 Daphna Oyserman and her coworkers: Oyserman, Bybee, and Terry, 2006.

145 Social psychologists Gregory Walton and Geoffrey Cohen: Walton and Cohen, 2007. See also another effective, brief intervention by G. L. Cohen, Garcia, Apfel, and Master (2006).

146 Herrnstein and Murray: Herrnstein and Murray, 1994.

146 Blacks start out high school: Myerson, Rank, Raines, and Schnitzler, 1998.

146   Psychologist Joel Myerson and his coworkers: Myerson, Rank, Raines, and Schnitzler, 1998.

147   A third possible answer: Aronson and Steele, 2005; Steele and Aronson, 1995.

147   One study that followed: Osborne, 1997.

148   Another long-term study: Massey and Fischer, 2005.

150   The Nobel Prize–winning economist James Heckman: Heckman, 2006.

150   The initial cost: Besharov, 2007.

150   The same is true: Masse and Barnett, 2002.

150   Even when benefits: Dickens and Baschnagel, 2008.

151   As a yardstick for measuring: Institute on Taxation and Economic Policy, 2007.

151   The cost per pupil: Institute of Education Sciences, 2006.

Chapter 8: Advantage Asia?

153   In 1966, Chinese Americans: Flynn, 1991a, 2007.

153   In 1980—when they were: Flynn, 2007.

153   In the late 1980s: Caplan, Whitmore, and Choy, 1989.

153   In 1999, U.S. eighth-graders: National Center for Education Statistics, 2000.

154   Although Asian Americans constitute: Quindlen, 2008a.

154   Asian and Asian American students: Anonymous, 2007; Anonymous, 2008a.

154   So European Americans: Throughout the discussion in this chapter, my generalizations are meant to apply only to East Asians and not to South Asians. I believe that some of the generalizations apply to South Asians as well, but the evidence—and there is a lot of it—concerns East Asians primarily. When I refer to "Asians" or "Easterners" in what follows I always mean East Asians such as Chinese, Japanese, Koreans, and citizens of Taiwan, Singapore, and Hong Kong, all of whom have cultures that are Confucian historically and at base today.

154   Herrnstein and Murray: Herrnstein and Murray, 1994.

154   Rushton and Jensen: Rushton and Jensen, 2005.

154   Philip Vernon: Vernon, 1982.

154   Richard Lynn: Lynn, 1987.

154   but Flynn: Flynn, 1991a.

154   Most showed that East Asians: All comparisons that I am aware of between Asians or Asian Americans on the one hand and European Americans on the other show that Asian and Asian American scores on performance IQ tests, and especially visuospatial tests such as block design, are high relative to scores on verbal IQ tests, though the difference is usually slight. Herrnstein and Murray attribute this relatively greater performance ability of Asians to genetic differences, which

certainly cannot be ruled out. On the other hand, as you will see later in this chapter, there are cultural differences that encourage attention to broad aspects of the visuospatial world more for Asians and Asian Americans. In addition, East Asian colleagues have told me that spatial problems resembling those found on IQ tests are taught in school to a much greater degree than is true in the West.

154   Harold Stevenson and his coworkers: Stevenson et al., 1990.

155   Even more astonishing: Stevenson and Stigler, 1992.

155   The Coleman report on educational quality: Flynn, 1991a.

156   Despite their slightly inferior: Flynn, 1991a.

156   By the age of thirty-two: Flynn, 1991a.

157   Recently, Flynn studied: Flynn, 2007.

157   And indeed they did: Flynn, 2007.

158   Asian and Asian American achievement: Nakanishi, 1982.

158   The high-school-age children: Caplan, Whitmore, and Choy, 1989.

158   Black eighth-grade children: Oyserman, Bybee, and Terry, 2006.

158   Asians today still believe: Chen and Stevenson, 1995; Choi and Markus, 1998; Choi, Nisbett, and Norenzayan, 1999; Heine et al., 2001; Holloway, 1988; Stevenson et al., 1990.

158   A team of Canadian psychologists: Heine et al., 2001.

159   But a still more important reason: Nisbett, 2003.

160   These East-West differences: My account of the social differences between Easterners and Westerners, as well as some of the resulting cognitive differences, is an abbreviation of points made in my book *The Geography of Thought: How Asians and Westerners Think Differently . . . and Why*. That book reports on historical and contemporary social and cognitive trends that differentiate Europeans and European Americans from East Asians and Americans of East Asian extraction. I refer you to that book for documentation of most of the assertions in this chapter about historical differences between Asians and Westerners and for detailed descriptions of a large number of studies on the perceptual and cognitive habits of contemporary people.

162   Takahiko Masuda and I: Masuda and Nisbett, 2001.

164   In another study, Masuda: Masuda, et al., 2008.

164   Chinese spend more time: Chua, Boland, and Nisbett, 2005.

164   Social psychologists have uncovered: Ross, 1977.

164   Koreans in this situation: Choi and Nisbett, 1998.

166   When we presented people: Ji, Zhang, and Nisbett, 2004.

166   We also presented: Norenzayan, Smith, Kim, and Nisbett, 2002.

166   My coworkers and I: Peng, 1997.

167   For example, when Chinese: Gutchess, Welsh, Boduroglu, and Park, 2006.

167   Consistent with this fact: Hedden, Ketay, Aron, Markus, and Gabrieli, 2008.

167   First, in several of the studies: Nisbett, 2003.

167   We found residents of Hong Kong: Ji, Zhang, and Nisbett, 2004.
167   And when Hong Kong residents: Hong, Chiu, and Kung, 1997.
168   The Japanese have spent: Organisation for Economic Co-operation and Development, 2007.
169   "I worked at the Carnegie Institution": French, 2001.
169   Second, the Confucian tradition: Munro, 1969; Nakamura, 1964/1985.
170   In fact, the children: Watanabe, 1998.
170   Nils Bohr credited: Bohr, 1958.

Chapter 9: People of the Book

171   "[The Jews] are peculiarly": Jewish Virtual Library, 2007.
171   "The United States today": Sugar, 2006.
171   American Jews have received: Anonymous, 2008b; JINFO.ORG, 2008.
172   Approximately the same overrepresentation: Anonymous, 2008b; JINFO.ORG, 2008.
172   And 26 to 34 percent: Anonymous, 2003; Anonymous, 2008b.
172   In the United States, Jews: Freedman, 2000.
172   an approximately equal: Anonymous, 2003.
172   and about 30 percent: Anonymous, 2003.
172   According to the 1931 census: Statistics in this paragraph come from Marcus (1983).
172   Most reports place: Backman, 1972; Lesser, Fifer, and Clark, 1965; Levinson, 1959; Majoribanks, 1972.
173   If we assume an average IQ of 110: Incidentally, despite this difference in the probability of having an IQ of 140, the probability of a random Jew having an IQ that is higher than that of the average white non-Jew is only .75, assuming an average Jewish IQ of 110; assuming an average Jewish IQ of 115, the probability is .84. Such is the surprising outcome predicted by the elevation of the normal distribution curve at various degrees of distance from the mean.
173   It is important to note: Burg and Belmont, 1990; Ortar, 1967; Patai, 1977. Before leaving the topic of Jewish IQ, I should note that there is an anomaly concerning Jewish intelligence. The major random samples of Americans having large numbers of Jewish participants show that whereas verbal and mathematical IQ run 10 to 15 points above the non-Jewish average, scores on tests requiring spatial-relations ability (ability to mentally manipulate objects in two- and three-dimensional space) are about 10 points below the non-Jewish average (Flynn, 1991a). This is an absolutely enormous discrepancy and I know of no ethnic group that comes close to having this 20 to 25-point difference among Jews. I do not for a minute doubt that the discrepancy is real.

I know half a dozen Jews who are at the top of their fields who are as likely to turn in the wrong direction as in the right direction when leaving a restaurant. The single ethnic difference that I believe is likely to have a genetic basis is the relative Jewish incapacity for spatial reasoning. I have no theory about why this should be the case, but I note that it casts an interesting light on the Jews' wandering in the desert for forty years!

173 This is true even: Gross, 1978.
174 Complete elimination from reproduction: Loehlin, Vandenberg, and Osborne, 1975.
174 The geneticist Cyril Darlington: Darlington, 1969.
175 Political scientist Charles Murray: Murray, 2007b.
176 Anthropologists Gregory Cochran, Jason Hardy: Cochran, Hardy, and Harpending, 2005.
178 Fifteen percent of all scientists: Sarton, 1975.
178 Even within Europe: In his book *Human Accomplishment* Charles Murray (2003) documented the swings of achievement in different regions.
179 "It is counted a title": Zweig, 1943/1987.
180 Psychologist Seymour Sarason: Sarason, 1973.

Chapter 10: Raising Your Child's Intelligence . . . and Your Own

183 There is no evidence: Bruer, 1999.
184 The research suggesting: Bruer, 1999.
184 We find the same kind: Duyme, 1981.
184 an exceptionally well-designed study: Clapp, 1996.
184 Exercising large muscle groups: G. D. Cohen, 2005.
184 Experiments show that elderly: Colcombe and Kramer, 2003.
185 You can even start: Aamodt and Wang, 2007.
185 Breast-feeding beyond nine months: Mortensen, Michaelsen, Sanders, and Reinisch, 2002.
185 It seems to be particularly important: Anderson, Johnstone, and Remley, 1999.
185 The activities that increase: Klingberg, Koenig, and Bilbe, 2002; Mortensen, Michaelsen, Sanders, and Reinish, Oleson, Westerberg, and Klingberg, 2003.
185 Neuroscientist Rosario Rueda: Rueda, Rothbart, McCandliss, Saccomanno, and Posner, 2005.
186 Child neurologist Torkel Klingberg: Klingberg, Koenig, and Bilbe, 2002.
186 Similar exercises improved: If your child has attention deficit hyperactivity disorder, you may want to contact a specialist skilled in admin-

istering training like that used by Klingberg and his colleagues (2002). The list is available at http://www.cogmed.com/cogmed/articles/en/78 .aspx

186 Finally, meditation exercises: Tang et al., 2007.

186 Personality psychologist Walter Mischel: Ayduk, Downey, Testa, Yen, and Shoda, 1999; Mischel, Shoda, and Rodriguez, 1989.

187 Recall that psychologists: Duckworth and Seligman, 2005.

187 We do know that if: Mischel, Shoda, and Rodriguez, 1989.

187 Also, Mischel and his coworkers had: Mischel, Shoda, and Rodriguez, 1989.

188 And Asians work harder: Heine et al., 2001.

189 When children are praised: Mueller and Dweck, 1998.

189 In a clever experiment: Mueller and Dweck, 1998.

190 With developmental psychologists Mark Lepper and David Green: Lepper, Greene, and Nisbett, 1973.

190 If a child has low: Calder and Staw, 1975; Loveland and Olley, 1979; V. C. McLoyd, 1979.

Epilogue: *What We Now Know about Intelligence and Academic Achievement*

195 First of all: Gottfredson, 1997.

195 Heuristics for reasoning: Nisbett, 1992.

195 Planning and choosing: Larrick, Morgan, and Nisbett, 1990; Nisbett, Fong, Lehman, and Cheng, 1987.

REFERENCES

Aamodt, S., and Wang, S. (2007, November 8). Exercise on the brain. *New York Times*. Retrieved November 8, 2007, from http://www.nytimes.com/2007/11/08/opinion/08aamodt.html?ref=opinion.

Adams, J., and Ward, R. H. (1973). Admixture studies and the detection of selection. *Science, 180*, 1137–1143.

Allington, R. L., and McGill-Franzen, A. (2003). Summer loss. *Phi Delta Kappan, 85*, 68–75.

Anderson, J. W., Johnstone, B. M., and Remley, D. T. (1999). Breast-feeding and cognitive development: A meta-analysis. *American Journal of Clinical Nutrition, 70*, 525–535.

Ankney, C. D. (1992). Sex differences in relative brain size: The mismeasure of woman, too. *Intelligence, 16*, 329–336.

Anonymous. (2003, September). Assessing the Ashkenazic IQ. *La Griffe du Lion, 5* (2). Retrieved April 1, 2008, from http://www.lagriffedulion.f2s.com/ashkenaz.htm.

Anonymous. (2007). A record pool leads to record results. *Harvard University Gazette*. Retrieved May 27, 2008, from http://www.news.harvard.edu/gazette/2007/04.05/99-admissions.html.

Anonymous. (2008a). Berkeley student protest to keep Asian study courses. *Sing Tao Daily*. Retrieved May 27, 2008, from http://news.newamericamedia.org/news/view_article.html?article_id=56e1b7053b6dc21ce22313 4d6e2531ed.

Anonymous. (2008b). Jewish Turing Mathematics Prizes, Fields Medal and others. *Israel Times*. Retrieved May 28, 2008, from http://www.israel-times.com/people/science-technology-nobels/jewish-turing-mathematics-prizes-fields-medal-and-others/.

Armor, D. (1976). *Analysis of the school preferred reading program in selected Los Angeles minority schools*. (Report No. R-2007-LAUSD). Santa Monica, CA: RAND.

Aronson, J., Fried, C. B., and Good, C. (2002). Reducing stereotype threat and boosting academic achievement of African-American students: The role of conceptions of intelligence. *Journal of Experimental Social Psychology, 38*, 113–125.

Aronson, J., and Steele, C. M. (2005). Stereotypes and the fragility of academic competence, motivation, and self-concept. In E. Elliot and C. Dweck (Eds.), *Handbook of competence and motivation*. New York: Guilford.

Attewell, P., Domina, T., Lavin, D., and Levey, T. (2004). The black middle class: Progress, prospects and puzzles. *Journal of African American Studies, 8*, 6–19.

Ayduk, O., Downey, G., Testa, A., Yen, Y., and Shoda, Y. (1999). Does rejection elicit hostility in high rejection sensitive women? *Social Cognition, 17*, 245–271.

Backman, M. E. (1972). Patterns of mental abilities: Ethnic, socioeconomic, and sex differences. *American Educational Research Journal, 9*, 1–12.

Baddeley, A. (1986). *Working memory*. Oxford: Oxford University Press/Clarendon Press.

Baghurst, P. A. (1992). Environmental exposure to lead and children's intelligence at the age of seven years: The Port Pirie cohort study. *New England Journal of Medicine, 327*, 1279–1284.

Bakalar, N. (2007). Study points to genetics in disparities in preterm births. *New York Times*. Retrieved June 29, 2008, from http://query.nytimes.com/gst/fullpage.html?res=9E01E5DC1E3EF934A15751C0A9619C8B63&sec=&spon=&pagewanted=all.

Baltes, P. B., and Reinert, G. (1969). Cohort effects in cognitive development in children as revealed by cross sectional sequences. *Developmental Psychology, 1*, 169–177.

Barnett, W. S. (1992). Benefits of compensatory preschool education. *Journal of Human Resources, 27*, 279–312.

Barnett, W. S. (2007). Commentary: Benefit-cost analysis of early childhood programs. *Social Policy Report, 21*, 12–13.

Bazelon, E. (2008). The next kind of integration [Electronic Version]. *New York Times*. Retrieved July 21, 2008, from http://www.nytimes.com/2008/07/20/magazine/20integration-t.html?pagewanted=1&sq=wake%20county&st=cse&scp=1.

Beals, K. L., Smith, C. L., and Dodd, S. M. (1984). Brain size, cranial morphology, climate and time machines. *Current Anthropology, 25*, 301–330.

Bergstrand, R., Vedin, A., Wilhelmsson, C., and Wilhelmsen, L. (1983). Bias due to non-participation and heterogeneous sub-groups in the population. *Journal of Chronic Diseases, 36*, 725–728.

Besharov, D. J. (2005). *Head Start's broken promise*. Washington, DC: American Enterprise Institute.

Besharov, D. J. (2007). *Testimony before the Joint Economic Committee, U.S. Congress. "Investing in young children pays dividends: The economic case for early care and education"*. Washington, DC: American Enterprise Institute for Public Policy Research.

Besharov, D. J., Germanis, P., and Higney, C. (2006). *Summaries of twenty early childhood evaluations.* College Park, MD: Maryland School of Public Affairs.

Bifulco, R., and Ladd, H. F. (2006). The impacts of charter schools on student achievement: Evidence from North Carolina. *Education Finance and Policy, 1,* 50–90.

Blackwell, L., Trzesniewski, K., and Dweck, C. S. (2007). Implicit theories of intelligence predict achievement across an adolescent transition: A longitudinal study and an intervention. *Child Development, 78,* 246–263.

Blair, C. (2006). How similar are fluid cognition and general intelligence? A developmental neuroscience perspective on fluid cognition as an aspect of human cognitive ability. *Behavioral and Brain Sciences, 29,* 109–160.

Blair, C., Gamson, D., Thorne, S., and Baker, D. (2005). Rising mean IQ: Cognitive demand of mathematics education for young children, population exposure to formal schooling, and the neurobiology of the prefrontal cortex. *Intelligence, 33,* 93–106.

Blair, C., and Razza, R. P. (2007). Relating effortful control, executive function, and false belief understanding to emerging math and literacy ability in kindergarten. *Child Development, 78,* 647–663.

Bohr, N. (1958). *Atomic physics and human knowledge.* New York: Wiley.

Borman, G. D., Hewes, G. M., Overman, L. T., and Brown, S. (2003). Comprehensive school reform and achievement: A meta-analysis. *Review of Educational Research, 73,* 125–230.

Borman, G. D., Slavin, R. E., Cheung, A., Chamberlain, A., Madden, N., and Chambers, B. (2007). *Final reading outcomes of the national randomized field trial of Success for All.* Madison: University of Wisconsin–Madison.

Bouchard, T. J. (1983). Do environmental similarities explain the similarity in intelligence of identical twins reared apart? *Intelligence, 7,* 175–184.

Bouchard, T. J. (2004). Genetic influence on human psychological traits. *Current Directions in Psychological Science, 13,* 148–151.

Bouchard, T. J., and McGue, M. (2003). Genetic and environmental influences on human psychological differences. *Journal of Neurobiology, 54,* 4–45.

Braver, T. S., and Barch, D. M. (2002). A theory of cognitive control, aging cognition, and neuromodulation. *Neuroscience and Biobehavioral Reviews, 26,* 809–817.

Bronfenbrenner, U. (1986). Ecology of the family as a context for human development: Research perspectives. *Developmental Psychology, 22,* 723–742.

Bronfenbrenner, U. (1975/1999). Nature with nurture: A reinterpretation of the evidence. In A. Montagu (Ed.), *Race and IQ* (2nd ed.). New York: Oxford University Press.

Brooks-Gunn, J., and Markman, L. B. (2005). The contribution of parenting to ethnic and racial gaps in school readiness. *Future of Children, 15,* 139–168.

Brooks-Gunn, J., McCarton, C. M., Casey, P. H., McCormick, M. C., Bauer, C. R., Bernbaum, J. C., et al. (1994). Early intervention in low birthweight, premature infants: Results through age 5 years from the Infant Health and Development Program. *Journal of the American Medical Association, 272,* 1257–1262.

Brown, P. (1992). Recent human evolution in East Asia and Australasia. *Philosophical Transaction of the Royal Society of London B, 337,* 235–242.

Brown, P., and Maeda, T. (2004). Post-Pleistocene diachronic change in East Asian facial skeletons: The size, shape and volume of the orbits. *Anthropological Science, 11,* 20–40.

Bruer, J. T. (1999). *The myth of the first three years: A new understanding of early brain development and lifelong learning.* New York: Free Press.

Burg, B., and Belmont, I. (1990). Mental abilities of children from different cultural backgrounds in Israel. *Journal of Cross-Cultural Psychology, 21,* 90–108.

Burkham, D. T., Ready, D. D., Lee, V. E., and LoGerfo, L. F. (2004). Social-class differences in summer learning between kindergarten and first grade: Model specification and estimation. *Sociology of Education, 77,* 1–31.

Burrell, B. (2005). *Postcards from the brain museum: The improbable search for meaning in the matter of famous minds.* New York: Broadway/Random House.

Burt, C., Jones, E., Miller, E., and Moodie, W. (1934). *How the mind works.* New York: Appleton-Century-Crofts.

Cahan, S., and Cohen, N. (1989). Age vs. schooling effects on intelligence development. *Child Development, 60,* 1239–1249.

Calder, B. J., and Staw, B. M. (1975). Self-perception of intrinsic and extrinsic motivation. *Journal of Personality and Social Psychology, 31,* 599–605.

Camarota, S. A. (2007). *Illegitimate nation: An examination of out-of-wedlock births across immigrants and natives.* Washington, DC: National Center for Health Statistics.

Campbell, F. A., Pungello, E. P., Miller-Johnson, S., Burchinal, M., et al. (2001). The development of cognitive and academic abilities: Growth curves from an early childhood educational experiment. *Developmental Psychology, 37,* 231–242.

Campbell, F. A., and Ramey, C. T. (1995). Cognitive and school outcomes for high-risk African-American students at middle adolescence: Positive effects of early intervention. *American Educational Research Journal, 32,* 743–772.

Caplan, N., Whitmore, J. K., and Choy, M. H. (1989). *The boat people and achievement in America: A study of economic and eductional success.* Ann Arbor: University of Michigan Press.

Capron, C., and Duyme, M. (1989). Assessment of the effects of socio-economic status on IQ in a full cross-fostering study. *Nature, 340,* 552–554.

Carter, S. C. (2000). *No excuses. Lessons from 21 high-performing, high-poverty schools.* Washington, DC: Heritage Foundation.

Caspi, A., Williams, B., Kim-Cohen, J., Craig, I. W., et al. (2007). Moderation of breastfeeding effects on the IQ by genetic variation in fatty acid metabolism. *Proceedings of the National Academy of Sciences of the United States of America, 104,* 18860.

Cattell, R. B. (1987). *Intelligence: Its structure, growth and action.* Amsterdam: North-Holland.

Cavanaugh, J. C., and Blanchard-Fields, F. (2006). *Adult development and aging* (5th ed.). Belmont, CA: Thomson Wadsworth.

Ceci, S. J. (1991). How much does schooling influence general intelligence and its cognitive components? A reassessment of the evidence. *Developmental Psychology, 27,* 703–722.

Ceci, S. J. (2007). *Racial, ethnic and socioeconomic achievement gaps: A cross-disciplinary critical review.* Ithaca, NY: Cornell University.

Ceci, S. J., and Williams, W. M. (1997). Schooling, intelligence and income. *American Psychologist, 52,* 1051–1058.

Centers for Disease Control and Prevention. (2007). Fetal alcohol spectrum disorders. Atlanta, GA. CDC. Retrieved January 19, 2008, from http://www.cdc.gov/ncbddd/fas/fasask.htm.

Chase-Lansdale, P., Mott, F. L., Brooks-Gunn, J., Phillips, D. A., et al. (1991). Children of the NLSY: A unique research opportunity. *Developmental Psychology, 27,* 918–931.

Chen, C., and Stevenson, H. W. (1995). Motivation and mathematics achievement: A comparative study of Asian-American, Caucasian-American and East Asian high school students. *Child Development, 66,* 1215–1234.

Choi, I., and Markus, H. R. (1998). *Implicit theories and causal attribution East and West.* Unpublished manuscript, Ann Arbor: University of Michigan.

Choi, I., and Nisbett, R. E. (1998). Situational salience and cultural differences in the correspondence bias and in the actor-observer bias. *Personality and Social Psychology Bulletin, 24,* 949–960.

Choi, I., Nisbett, R. E., and Norenzayan, A. (1999). Causal attribution across cultures: Variation and universality. *Psychological Bulletin, 125,* 47–63.

Chua, H. F., Boland, J. E., and Nisbett, R. E. (2005). Cultural variation in eye movements during scene perception. *Proceedings of the National Academy of Sciences of the United States of America, 102,* 12629–12633.

Churchill, W. (1974). *A history of the English-speaking peoples.* New York: Bantam.

Clapp, J. F. (1996). Morphometric and neurodevelopmental outcome at age five years of the offspring of women who continued to exercise regularly throughout pregnancy. *Journal of Pediatrics, 129,* 856–863.

Clapp, J. F., Kim, H., Burciu, B., and Lopez, B. (2000). Beginning regular exercise in early pregnancy: Effect on fetoplacental growth. *American Journal of Obstetrics and Gynecology, 183,* 1484–1488.

Cochran, G., Hardy, J., and Harpending, H. (2005). Natural history of Ashkenazi intelligence. *Journal of Biosocial Science, 38,* 1–35.

Cohen, G. D. (2005). *The mature mind: The positive power of the aging brain.* New York: Basic Books.

Cohen, G. L., Garcia, J., Apfel, N., and Master, A. (2006). Reducing the racial achievement gap: A social-psychological intervention. *Science, 313,* 1307–1310.

Colcombe, S., and Kramer, A. F. (2003). Fitness effects on the cognitive function of older adults: A meta-analytic study. *Psychological Science, 14,* 125–130.

Connell, N. (1996). *Getting off the list: School improvement in New York City.* New York: Robert Sterling Clark Foundation.

Cook, T. D. (2003). Why have educational evaluators chosen not to do randomized experiments? *Annals, American Academy of Political and Social Science, 589,* 114–149.

Cook, T. D., Habib, F., Phillips, M., Settersen, R. A. et al. (1999). Comer's school development program in Prince George's County Maryland: A theory-based evaluation. *American Educational Research Journal, 36,* 543–597.

Cook, T. D., Hunt, H. D., and Murphy, R. F. (2000). Comer's school development program in Chicago: A theory-based evaluation. *American Educational Research Journal, 37,* 535–597.

Cooper, H., Nye, B., Charlton, K., Lindsay, J., and Greathouse, S. (1996). The effects of summer vacation on achievement test scores: A narrative and meta-analytic review. *Review of Educational Research, 66,* 227–268.

Daley, T. C., Whaley, S. E., Sigman, M. D., Espinosa, M. P., and Neumann, C. (2003). IQ on the rise: The Flynn effect in rural Kenyan children. *Psychological Science, 14,* 215–219.

Darity, W. A., Jr., and Mason, P. L. (1998). Evidence on discrimination in employment: Codes of color, codes of gender. *Journal of Economic Perspectives, 12,* 63–90.

Darley, J. M., and Berscheid, E. (1967). Increased liking as a result of the anticipation of personal contact. *Human Relations, 20,* 29–40.

Darlington, C. (1969). *The evolution of man and society.* London: Allen and Unwin.

David, J. L., Woodworth, K., Grant, E., Guha, R., Lopez-Torkos, A., and Young, V. M. (2006). *Bay Area KIPP Schools: A study of early implementation.* Menlo Park, CA: SRI International.

De Sander, M. K. (2000). Teacher evaluation and merit pay: Legal considerations, practical concerns. *Journal of Personnel Evaluation in Education, 14,* 301–317.

Deary, I. J. (2001). *Intelligence: A very short introduction.* New York: Oxford University Press.

DeGroot, A. D. (1948). The effects of war upon the intelligence of youth. *Journal of Abnormal and Social Psychology, 43*, 311–317.

Deming, D. (2008). *Early childhood intervention and life-cycle skill development*. Cambridge, MA: Harvard University.

Der, G., Batty, G. D., & Deary, I. J. (2006). Effect of breast feeding on intelligence in children: Prospective study, sibling pairs analysis, and meta-analysis. *British Medical Journal, 333*, 945–948.

Devlin, B., Daniels, M., and Roeder, K. (1997). The heritability of IQ. *Nature, 388*, 468–471.

Diamond, A., Barnett, W. S., Thomas, J., and Munro, S. (2007). Preschool program improves cognitive control. *Science, 318*, 1387–1338.

Dickens, W. T., and Baschnagel, C. (2008). *Dynamic estimates of the fiscal effects of investing in early childhood programs*. Washington, DC: Brookings Institution.

Dickens, W. T., and Flynn, J. R. (2001). Heritability estimates versus large environmental effects: The IQ paradox resolved. *Psychological Review, 108*, 346–369.

Dickens, W. T., and Flynn, J. R. (2006). Black Americans reduce the racial IQ gap: Evidence from standardization samples. *Psychological Science, 17*, 913–920.

Dillman, D. A. (1978). *Mail and telephone surveys: The Total Design Method*. New York: John Wiley and Sons.

Dodge, K. A., Pettit, G., & Bates, J. (1994). Socialization mediators of the relation between socioeconomic status and child conduct problems. *Child Development, 62*, 583–599.

Dotinga, A., Schrijvers, C. T. M., Voorham, A. J. J., and Mackenbach, J. P. (2005). Correlates of stages of change of smoking among inhabitants of deprived neighborhoods. *European Journal of Health, 15*, 152–159.

Duckworth, A. L., and Seligman, M. E. P. (2005). Self-discipline outdoes IQ in predicting academic performance of adolescents. *Psychological Science, 16*, 939–944.

Duncan, J., Burgess, P., and Emslie, H. (1995). Fluid intelligence after frontal lobe lesions. *Neuropsychologia, 33*, 261–268.

Duyme, M. (1981). *Les enfants abandonnés. Rôle des familles adoptives et des assistantes maternelles*. Paris: CNRS.

Duyme, M., Dumaret, A., and Tomkiewicz, S. (1999). How can we boost IQs of "dull" children? A late adoption study. *Proceedings of the National Academy of Sciences of the United States of America, 96*, 8790–8794.

Eicholz, R. (1991). *Addison-Wesley Mathematics: Grade 2*. Atlanta: Pearson Education.

Evers, W. M., and Clopton, P. (2006). High-spending, low-performing school districts. In E. A. Hanushek (Ed.), *Courting failure: How school finance lawsuits exploit judges' good intentions and harm our children*. Stanford, CA: Education Next Books.

Eyferth, K. (1961). Leistungern verschiedener Gruppen von Besatzungskindern in Hamburg-Wechsler Intelligenztest für Kinder (HAWIK). *Archiv für die gesamte Psychologie, 113*, 222–241.

Eysenck, H. J. (1971). *The IQ argument: Race, intelligence and education.* New York: Library Press.

Eysenck, H. J. (1973). *The inequality of man.* London: Temple Smith.

Fagan, J. F., and Holland, C. R. (2002). Equal opportunity and racial differences in IQ. *Intelligence, 30*, 361–387.

Fagan, J. F., and Holland, C. R. (2007). Racial equality in intelligence: Predictions from a theory of intelligence as processing. *Intelligence, 35*, 319–334.

Feldman, M. W., and Otto, S. P. (1997). Twin studies, heritability, and intelligence. *Science, 278*, 1384–1385.

Finder, A. (2005, September 25). As test scores jump, Raleigh credits integration by income. *New York Times*, p. 1.

Fischbein, S. (1980). IQ and social class. *Intelligence, 4*, 51–63.

Flynn, J. R. (1980). *Race, IQ and Jensen.* London: Routledge and Kegan Paul.

Flynn, J. R. (1987). Massive IQ gains in 14 nations: What IQ tests really measure. *Psychological Bulletin, 101*, 171–191.

Flynn, J. R. (1991a). *Asian Americans: Achievement beyond IQ.* Hillsdale, NJ: Lawrence Erlbaum.

Flynn, J. R. (1991b). Reaction times show that both Chinese and British children are more intelligent than one another. *Perceptual and Motor Skills, 72*, 544–546.

Flynn, J. R. (1994). IQ gains over time. In R. J. Sternberg (Ed.), *The encyclopedia of human intelligence* (pp. 617–623). New York: Macmillan.

Flynn, J. R. (1998). IQ gains over time. In U. Neisser (Ed.), *The rising curve: Long term gains in IQ and related measures* (pp. 25–66). Washington, DC: American Psychological Association.

Flynn, J. R. (2000a). IQ gains, WISC subtests and fluid *g: g* theory and the relevance of Spearman's hypothesis to race. In G. R. Bock, J. Goode, and K. Webb (Eds.), *The nature of intelligence.* New York: Wiley.

Flynn, J. R. (2000b). IQ trends over time: Intelligence, race, and meritocracy. In K. Arrow, S. Bowles, and S. Durlauf (Eds.), *Meritocracy and economic inequality.* Princeton, NJ: Princeton University Press.

Flynn, J. R. (2007). *What is intelligence? Beyond the Flynn effect.* New York: Cambridge University Press.

Flynn, J. R. (2008). *Where have all the liberals gone? Race, class, and ideals in America.* New York: Cambridge University Press.

Folger, J. K., and Nam, C. B. (1967). *Education of the American population* (A 1960 U.S. Census monograph). Washington, DC: U.S. Department of Commerce.

Freedman, J. O. (2000). Ghosts of the past: Anti-Semitism at elite colleges. *Chronicle of Higher Education, 47*(4).

Freeman, F. S. (1934). *Individual differences: The nature and causes of variations in intelligence and special abilities.* New York: Holt.

Freeman, R. B. (2007). *America works: The exceptional U.S. labor market.* New York: Russell Sage Foundation.

French, H. W. (2001, August 7). Hypothesis: A scientific gap. Conclusion: Japanese custom. *New York Times,* p. A1.

General Accounting Office. (1999). *Lead poisoning: Federal health-care programs are not effectively reaching at-risk children.* Washington, DC: General Accounting Office.

Garber, H. L. (1988). *The Milwaukee Project: Preventing mental retardation in children at risk.* Washington, DC: American Association on Mental Retardation.

Garces, E., Thomas, D., and Currie, J. (2002). Longer-term effects of Head Start. *American Economic Review, 92,* 999–1012.

Gardner, H. (1983/1993). *Frames of mind: The theory of multiple intelligences.* New York: Basic Books.

Good, C., Aronson, J., and Inzlicht, M. (2003). Improving adolescents' standardized test performance: An intervention to reduce the effects of stereotype threat. *Applied Developmental Psychology, 24,* 645–662.

Gormley, W. T., Jr., Gayer, T., Phillips, D., and Dawson, B. (2005). The effects of universal pre-K on cognitive development. *Developmental Psychology, 41,* 872–884.

Gottfredson, L. S. (1997). Intelligence and social policy. *Intelligence, 24,* 1–320.

Gould, S. J. (1981). *The mismeasure of man.* New York: W. W. Norton.

Gray, J. R., and Thompson, P. M. (2004). Neurobiology of intelligence: Science and ethics. *Nature Reviews: Neuroscience, 5,* 471–482.

Green, C. S., and Bavelier, D. (2003). Action video game modifies visual selective attention. *Nature, 423,* 534–537.

Green, R. L., Hoffman, L. T., Morse, R., Hayes, M. E., and Morgan, R. F. (1964). *The educational status of children in a district without public schools* (Co-Operative Research Project No. 2321). Washington, DC: Office of Education, U.S. Department of Health, Education and Welfare.

Grissmer, D., Flanagan, A., and Williamson, S. (1998). Why did the black-white score gap narrow in the 1970s and 1980s? In C. Jencks and M. Phillips (Eds.), *The black-white test score gap.* Washington, DC: Brookings Institution Press.

Gross, M. B. (1978). Cultural concomitants of preschoolers' preparation for learning. *Psychological Reports, 43,* 807–813.

Gross, R. T., Spiker, D., and Haynes, C. W. (1997). *Helping low birth weight, premature babies: The Infant Health and Development Program.* Stanford, CA: Stanford University Press.

Guevara-Aguire, J., Rosenbloom, A. L., Vaccarelo, M. A., Fielder, P. J., de la Vega, A., Diamond, F. B., et al. (1991). Growth hormone receptor deficiency (Laron syndrome): Clinical and genetic characteristics. *Acta Paediatrica Scandinavia, 377* (Suppl.), 96–103.

Gutchess, A. H., Welsh, R. C., Boduroglu, A., and Park, D. C. (2006). Cultural differences in neural function associated with object processing. *Cognitive, Affective and Behavioral Neuroscience, 6,* 102–109.

Hack, M., Klein, N., and Taylor, H. G. (1995). Long-term developmental outcomes of low birth weight infants. *Future of Children, 5,* 176–196.

Hamre, B. K., and Pianta, R. C. (2001). Early teacher-child relationships and the trajectory of children's school outcomes through eighth grade. *Child Development, 72,* 625–638.

Hamre, B. K., and Pianta, R. C. (2005). Can instructional and emotional support in the first-grade classroom make a difference for children at risk of school failure? *Child Development, 76,* 949–967.

Hanushek, E. A. (1986). The economics of schooling: Production and efficiency in public schools. *Journal of Economic Literature, 24,* 1141–1177.

Hanushek, E. A. (2002). *The failure of input-based schooling policies* (Working Paper No. 9040). Cambridge, MA: National Bureau of Education Research.

Hanushek, E. A., Kain, J. F., O'Brien, D. M., and Rivkin, S. G. (2005). *The market for teacher quality* (Working Paper No. 11154). Cambridge, MA: National Bureau of Economic Research.

Harden, K. P., Turkheimer, E., and Loehlin, J. C. (2006). Genotype by environment interaction in adolescents' cognitive aptitude. *Behavior Genetics, 37,* 273–283.

Härnqvist, K. (1968). Changes in intelligence from 13 to 18. *Scandinavian Journal of Psychology, 9,* 50–82.

Harris, J. R. (1998). *The nurture assumption: Why children turn out the way they do.* New York: Touchstone.

Hart, B., and Risley, T. (1995). *Meaningful differences in the everyday experience of young American children.* Baltimore: Brookes.

Hayes, D., & Grether, J. (1983). The school year and vacations: When do students learn? *Cornell Journal of Social Relations, 17,* 56-71.

Heath, S. B. (1982). What no bedtime story means: Narrative skills at home and school. *Language in Society, 11,* 49–79.

Heath, S. B. (1983). *Ways with words: Language, life, and work in communities and classrooms.* Cambridge: Cambridge University Press.

Heath, S. B. (1990). The children of Trackton's children. In J. W. Stigler, R. A. Shweder, and G. Herdt (Eds.), *Cultural psychology: Essays on comparative human development.* Cambridge: Cambridge University Press.

Heckman, J. J. (2006). Skill formation and the economics of investing in disadvantaged children. *Science, 312,* 1900–1902.

Hedden, T., Ketay, S., Aron, A., Markus, H. R., and Gabrieli, J. D. (2008). Cultural influences on neural substrates of attentional control. *Psychological Science, 19,* 12–17.

Heine, S. J., Kitayama, S., Lehman, D. R., Takata, T., Ide, E., Leung, C., et al. (2001). Divergent consequences of success and failure in Japan and North America: An investigation of self-improving motivation. *Journal of Personality and Social Psychology, 81*, 599–615.

Henderson, V. L., and Dweck, C. S. (1990). Achievement and motivation in adolescence: A new model and data. In S. Feldman and G. Elliott (Eds.), *At the threshold: The developing adolescent.* Cambridge, MA: Harvard University Press.

Henneberg, M. (1988). Brain size/body weight variability in modern humans: Consequences for interpretations of hominid evolution. *South African Journal of Science, 84*, 521–522.

Henneberg, M., and Steyn, M. (1993). Trends in cranial capacity and cranial index in subsaharan Africa during the Holocene. *American Journal of Human Biology, 5*, 473–479.

Henneberg, M., and Steyn, M. (1995). Diachronic variation of cranial size and shape in the Holocene: A manifestation of hormonal evolution? *Rivista di Antropologia, 73*, 159–164.

Herrnstein, R. J., and Murray, C. (1994). *The bell curve: Intelligence and class structure in American life.* New York: Free Press.

Herrnstein, R. J., Nickerson, R. S., Sanchez, M., and Swets, J. A. (1986). Teaching thinking skills. *American Psychologist, 41*, 1279-1289.

Hess, F. M. (2006). *Stimulant or slave? The politics of adequacy implementation.* Paper presented at annual meeting of the American Political Science Association, Philadelphia.

Hill, J. L., Brooks-Gunn, J., and Waldfogel, J. (2003). Sustained effects of high participation in an early intervention for low-birth-weight premature infants. *Developmental Psychology, 39*, 730–744.

Ho, K. C., Roessmann, U., Hause, L., and Monroe, G. (1981). Newborn brain weight in relation to maturity, sex, and race. *Annals of Neurology, 10*, 243–246.

Ho, K. C., Roessmann, U., Straumfjord, J. V., and Monroe, G. (1980). Analysis of brain weight: I and II. *Archives of Pathology and Laboratory Medicine, 104*, 635–645.

Holloway, S. (1988). Concepts of ability and effort in Japan and the United States. *Review of Educational Research, 58*, 327–345.

Hong, Y., Chiu, C., and Kung, T. (1997). Bringing culture out in front: Effects of cultural meaning system activation on social cognition. In K. Leung, Y. Kashima, U. Kim, and S. Yamaguchi (Eds.), *Progress in Asian social psychology* (Vol. 1, pp. 135–146). Singapore: Wiley.

Howell, W., Wolf, P., Peterson, P., and Campbell, D. (2001, Winter). Reply to Krueger. *Education Next, 5.*

Hoxby, C. M. (2000). The effects of class size on student achievement: New evidence from population variation. *Quarterly Journal of Economics, 115*, 1239–1285.

Hoxby, C. M. (2004). *Achievement in charter schools and regular public* schools in the United States: Understanding the differences. Cambridge, MA: Harvard University Press.

Hoxby, C. M., and Murarka, S. (2007). New York City's charter schools overall report. Cambridge, MA: New York City Charter Schools Evaluation Project.

Hoxby, C. M., and Rockoff, J. E. (2004). The impact of charter schools on student achievement. Cambridge, MA: Harvard University Press.

Husén, T. (1951). The influence of schooling upon IQ. Theoria, 17, 61–88.

Ignatiev, N. (1995). How the Irish became white. New York: Routledge.

Infant Health and Development Program. (1990). Enhancing the outcomes of low-birth-weight, premature infants: A multisite randomized trial. Journal of the American Medical Association, 263, 3035–3042.

Institute of Education Sciences. (2006). Digest of Education Statistics: 2005. Retrieved August 1, 2007, from http://nces.ed.gov/programs/digest/d05/tables_2.asp#Ch2Sub9.

Institute on Taxation and Economic Policy. (2007). The Bush tax cuts: The latest CTJ data March 2007. Washington, DC: Institute for Taxation and Economic Policy.

Jacob, B. A., and Lefgren, L. (2005). Principals as agents: Subjective performance measurement in education (Working Paper No. 11463). Cambridge, MA: National Bureau of Economic Research.

Jaeggi, S. M., Perrig, W. J., Jonides, J., and Buschkuehl, M. (2008). Improving fluid intelligence with training on working memory. Proceedings of the National Academy of Science of the United States of America, 105, 6829–6833.

Jencks, C., Smith, M., Acland, H., Bane, M. J., Cohen, D., Gintis, H., et al. (1972). Inequality: A reassessment of the effects of family and schooling in America. New York: Harper and Row.

Jensen, A. R. (1969, Winter). How much can we boost I.Q. and scholastic achievement? Harvard Educational Review, 1–123.

Jensen, A. R. (1980). Bias in mental testing. New York: Free Press.

Jensen, A. R. (1997). Adoption data and two g-related hypotheses. Intelligence, 25, 1–6.

Jensen, A. R. (1998). The g factor. Westport, CT: Praeger.

Jensen, A. R., and Whang, P. A. (1993). Reaction times and intelligence: A comparison of Chinese-American and Anglo-American children. Journal of Biosocial Science, 25, 397–410.

Jerald, C. (2001). Dispelling the myth revisited: Preliminary findings from a nationwide analysis of "high-flying" schools. Washington, DC: Education Trust.

Jessness, J. (2002). The untold story behind the famous rise—and shameful fall—of Jaime Escalante, America's master math teacher. Reason. Retrieved July 2, 2002, from http://www.reason.com/news/show/28479.html.

Jester, J. M., Nigg, J. T., Zucker, R. A., Puttler, L. I., Long, J. C., and Fitzgerald, H. E. (2008). Intergenerational transmission of neuropsy-

chological executive functioning. Unpublished manuscript, Ann Arbor: University of Michigan.

Jewish Virtual Library. (2007). *Mark Twain and the Jews.* Retrieved February 1, 2008, from http://www.jewishvirtuallibrary.org/jsource/US-Israel/twain.html.

Ji, L.-J., Zhang, Z., and Nisbett, R. E. (2004). Is it culture or is it language? Examination of language effects in cross-cultural research on categorization. *Journal of Personality and Social Psychology, 87,* 57–65.

JINFO.ORG. (2008). Retrieved April 1, 2008, from http://www.jinfo.org/Nobel_Prizes.html.

Johnson, S. (2005). *Everything bad is good for you: How today's popular culture is actually making us smarter.* New York: Riverhead Books.

Joiner, T. E. (in press). Head size as an explanation of the race-measured IQ relation: Negative evidence from child and adolescent samples. *Scientific Review of Mental Health Practice.*

Jooste, P. L., Yach, D., Steenkamp, H. J., and Rossouw, J. E. (1990). Dropout and newcomer bias in a community cardiovascular follow-up. *International Journal of Epidemiology, 19,* 284–289.

Juffer, F., Hoksbergen, R. A. C., Riksen-Walraven, J. M., and Kohnstamm, G. A. (1997). Early intervention in adoptive families: Supporting maternal sensitive responsiveness, infant-mother attachment, and infant competence. *Journal of Child Psychological Psychiatry, 38,* 1039–1050.

Kane, M. J., and Engle, R. W. (2002). The role of prefrontal cortex in working memory capacity, executive attention, and general fluid intelligence. *Psychonomic Bulletin and Review, 9,* 637–671.

Kane, T. (2007, June). *New findings on the effectiveness of National Board certified teachers and some implications for equity.* Paper presented at the Achievement Gap Conference, Cambridge, MA.

Kazui, H., Kitagaki, H., and Mori, E. (2000). Cortical activation during retrieval of arithmetical facts and actual calculation: A functional magnetic resonance imaging study. *Psychiatry and Clinical Neurosciences, 54,* 485.

Klingberg, A. K., Keonig, J. I., and Bilbe, G. (2002). Training of working memory in children with ADHD. *Journal of Clinical and Experimental Neuropsychology, 24,* 781–791.

Knudsen, E. I., Heckman, J. J., Cameron, J. L., and Shonkoff, J. P. (2006). Economic, neurobiological, and behavioral perspectives on building America's future workforce. *Proceedings of the National Academy of Sciences of the United States of America, 103,* 10155–10162.

Kramer, M. S. (2008). Breastfeeding and child cognitive development. *Archives of General Psychiatry, 65,* 578–584.

Kranzler, J. H., Rosenbloom, A. L., Martinez, V., and Guevara-Aguire, J. (1998). Normal intelligence with severe insulin-like growth factor I deficiency due to growth hormone receptor deficiency: A controlled study in a genetically homogenous population. *Journal of Clinical Endocrinology and Metabolism, 83,* 1953–1958.

Krueger, A. (1999). Experimental estimates of education production functions. *Quarterly Journal of Economics, 114*, 497–532.

Krueger, A. (2001, Winter). Vouchers versus class size. *Education Next,* 4–5.

Krueger, A., and Zhu, P. (2004). Another look at the New York City School Voucher Experiment. *American Behavioral Scientist, 47*, 658–698.

Kulik, J. (2003). *Effects of using instructional technology in elementary and secondary schools: What controlled evaluation studies say* (SRI Project No. P10446.001). Arlington, VA: SRI International.

Ladd, H. (2002). School vouchers: a critical view. *Journal of Economics, 16*, 3–24.

Landry, S. H., Smith, K. E., and Swank, P. R. (2006). Responsive parenting: Establishing early foundations for social, communication, and independent problem-solving skills. *Developmental Psychology, 42*, 627–642.

Landry, S. H., Smith, K. E., Swank, P. R., and Guttentag, C. (2007). *A responsive parenting intervention: The optimal timing across early childhood for impacting maternal behaviors and child outcomes.* Houston: University of Texas Health Science Center.

Lareau, A. (2003). *Unequal childhoods: Class, race, and family life.* Berkeley: University of California Press.

Larrick, R. P., Morgan, J. N., and Nisbett, R. E. (1990). Teaching the use of cost-benefit reasoning in everyday life. *Psychological Science, 1*, 362–370.

Lavy, V. (2002). Evaluating the effect of teacher performance incentives on students' achievements. *Journal of Political Economy, 110*, 1286–1317.

Lepper, M. R., Drake, M. F., and O'Donnell-Johnson, T. (1997). Scaffolding techniques of expert human tutors. In K. Hogan and M. Pressley (Eds.), *Scaffolding student learning: Instructional approaches and issues.* Cambridge, MA: Brookline Books.

Lepper, M. R., Greene, D., and Nisbett, R. E. (1973). Undermining children's intrinsic interest with extrinsic reward: A test of the overjustification hypothesis. *Journal of Personality and Social Psychology, 28*, 129–137.

Lepper, M. R., Wolverton, M., Mumme, D. L., and Gurtner, J.-L. (1993). Motivational techniques of expert human tutors: Lessons for the design of computer-based tutors. In S. P. Lajoie and S. J. Derry (Eds.), *Computers as cognitive tools.* Hillsdale, NJ: Lawrence Erlbaum.

Lepper, M. R., and Wolverton, M. (2001). The wisdom of practice: Lessons learned from the study of highly effective tutors. In J. Aronson (Ed.), *Improving academic achievement: Contributions of social psychology.* Orlando, FL: Academic Press.

Lesser, G. S., Fifer, G., and Clark, D. H. (1965). Mental abilities of children from different social-class and cultural groups. *Monographs of the Society for Research in Child Development, 30*, 1–115.

Levinson, B. (1959). A comparison of the performance of monolingual and bilingual native-born Jewish preschool children of traditional parentage on four intelligence tests. *Journal of Clinical Psychology, 15*, 74–76.

Levitt, S. D., and Dubner, S. J. (2006). *Freakonomics: A rogue economist explores the hidden side of everything.* New York: William Morrow.

Locurto, C. (1990). The malleability of IQ as judged from adoption studies. *Intelligence, 14,* 275–292.

Loehlin, J. C., Lindzey, G., and Spuhler, J. N. (1975). *Race differences in intelligence.* San Francisco: W. H. Freeman.

Loehlin, J. C., Vandenberg, S. G., and Osborne, R. T. (1973). Blood-group genes and Negro-white ability differences. *Behavior Genetics, 3,* 263–270.

Lopes, P. N., Grewal, D., Kadis, J., Gall, M., and Salovey, P. (2006). Evidence that emotional intelligence is related to job performance and affect and attitudes at work. *Psicothema, 18,* 132-138.

Love, J. M., Kisker, E. E., Ross, C., Raikes, H., et al. (2005). The effectiveness of Early Head Start for 3-year-old children and their parents: Lessons for policy and programs. *Developmental Psychology, 41,* 885–901.

Loveland, K. K., and Olley, J. G. (1979). The effect of external reward on interest and quality of task performance in children of high and low intrinsic motivation. *Child Development, 50,* 1207–1210.

Luca, A., Morley, R., Cole, T. J., Lister, G., and Leeson-Payne, C. (1992). Breast milk and subsequent intelligence quotient in children born preterm. *Lancet, 339,* 261–264.

Ludwig, J., and Miller, D. L. (2005). *Does Head Start improve children's life chances? Evidence from a regression discontinuity design* (Working Paper No. 11702). Cambridge, MA: National Bureau of Economic Research.

Ludwig, J., and Phillips, D. A. (2007). *The benefits and costs of Head Start* (Working Paper No. 12973). Cambridge, MA: National Bureau of Economic Research.

Lynn, R. (1987). The intelligence of the Mongoloids: A psychometric, evolutionary and neurological theory. *Personality and Individual Differences, 8,* 813–844.

Lynn, R., and Shighesia, T. (1991). Reaction times and intelligence: A comparison of Japanese and British children. *Journal of Biosocial Science, 23,* 409–416.

Lynn, R., and Vanhanen, T. (2002). *IQ and the wealth of nations.* Westport, CT: Praeger.

Macnamara, J. (1966). *Bilingualism and primary education: A study of Irish experience.* Edinburgh: Edinburgh University Press.

Majoribanks, K. (1972). Ethnic and environmental influences on mental abilities. *American Journal of Sociology, 78,* 323–337.

Marcus, J. (1983). *Social and political history of the Jews in Poland, 1919–1939.* Berlin: Mouton.

Masse, L. N., and Barnett, W. S. (2002). *A benefit cost analysis of the Abecedarian Early Childhood intervention.* New Brunswick, NJ: National Institute for Early Education Research.

Massey, D. S., & Fischer, M. J. (2005). Stereotype threat and academic performance: New data from the national survey of freshmen. *The Dubois Review: Social Science Research on Race, 2,* 45–68.

Masuda, T., Ellsworth, P. C., Mesquita, B., Leu, J., Tanida, S., and van de Veerdonk, E. (2008). Placing the face in context: Cultural differences in the perception of facial emotion. *Journal of Personality and Social Psychology, 94,* 365–381.

Masuda, T., and Nisbett, R. E. (2001). Attending holistically vs. analytically: Comparing the context sensitivity of Japanese and Americans. *Journal of Personality and Social Psychology, 81,* 922–934.

Mathews, J. (2006, January 17). America's best schools? *Washington Post.*

Maughan, B., and Collishaw, S. (1998). School achievement and adult qualifications among adoptees: A longitudinal study. *Journal of Child Psychology and Psychiatry and Allied Disciplines, 39,* 669–685.

McDaniel, M. A. (2005). Big-brained people are smarter: A meta-analysis of the relationship between in vivo brain volume and intelligence. *Intelligence, 33,* 337–346.

McFie, J. (1961). The effect of education on African performance on a group of intellectual tests. *British Journal of Educational Psychology, 31,* 232–240.

McGue, M., and Bouchard, T. J. (1998). Genetic and environmental influences on human behavioral differences. *Annual Review of Neuroscience, 21,* 1–24.

McGue, M., Bouchard, T. J., Iacono, W. G., and Lykken, D. T. (1993). Behavioral genetics of cognitive ability: A life-span perspective. In R. Plomin and G. E. McClearn (Eds.), *Nature, nurture and psychology.* Washington, DC: American Psychological Association.

McGue, M., Keyes, M., Sharma, A., Elkins, I., Legrand, L., Johnson, W., et al. (2007). The environments of adopted and non-adopted youth: Evidence on range restriction from the Sibling Interaction and Behavior Study (SIBS). *Behavior Genetics, 37,* 449–462.

McKey, R. H., Condelli, L., Ganson, B. B., McConkey, C., and Plantz, M. (1985). *The impact of Head Start on children, families and communities* (Final report of the Head Start Evaluation, Synthesis and Utilization Project.) Washington, DC: Department of Health and Human Services.

McLoyd, V. C. (1979). The effects of extrinsic rewards of differential value on high and low intrinsic interest. *Child Development, 50,* 1010–1019.

McLoyd, V. C. (1998). Socioeconomic disadvantage and child development. *American Psychologist, 53,* 185–204.

Meisenberg, G., Lawless, E., Lambert, E., and Newton, A. (2005). The Flynn effect in the Caribbean: Generational change in test performance in Dominica. *Mankind Quarterly, 46,* 29–70.

Mekel-Bobrov, N., et al. (2005). Ongoing adaptive evolution of *ASPM,* a brain size determinant in *Homo sapiens. Science, 309,* 1720–1722.

Micklewright, J., and Schnepf, S. V. (2004). *Educational achievement in English-speaking countries: Do different surveys tell the same story?*

Retrieved September 5, 2007, from ftp://repec.iza.org/RePEc/Discussion paper/dp1186.pdf.

Mikulecky, L. (1996). *Family literacy: Parent and child interactions.* Washington, D.C: U.S. Department of Education. Retrieved October 25, 2005, from http://www.ed.gov/pubs/FamLit/parent.html.

Mills, R. J., and Bhandari, S. (2003). *Health insurance coverage in the United States: 2002.* Washington, DC: U.S. Census Bureau.

Mischel, W. (1974). Processes in delay of gratification. In L. Berkowitz (Ed.), *Advances in experimental social psychology* (Vol. 7, pp. 249–292). New York: Academic Press.

Mischel, W., Shoda, Y., and Peake, P. K. (1988). The nature of adolescent competencies predicted by preschool delay of gratification. *Journal of Personality and Social Psychology, 54,* 687-696.

Mischel, W., Shoda, Y., and Rodriguez, M. L. (1989). Delay of gratification in children. *Science, 244,* 933-938.

Moore, E. G. J. (1986). Family socialization and the IQ test performance of traditionally and trans-racially adopted children. *Developmental Psychology, 22,* 317-326.

Mortensen, E. L., Michaelsen, K. M., Sanders, S. A., and Reinisch, J. M. (2002). The association between duration of breastfeeding and adult intelligence. *Journal of the American Medical Association, 287,* 2365-2371.

Moss, P., and Tilly, C. (2001). *Stories employers tell: Race, skill and hiring in America.* New York: Russell Sage Foundation.

Mosteller, F., and Boruch, R. (2002). *Evidence matters: Randomized trials in educational research.* Washington, DC: Brookings Institution.

Moynihan, D. P. (1965). *The Negro family: The case for national action.* Washington, DC: Government Printing Office.

Mueller, C. W., and Dweck, C. S. (1998). Praise for intelligence can undermine children's motivation and performance. *Journal of Personality and Social Psychology, 75,* 33-52.

Muijs, D., Harris, A., Chapman, C., Stoll, L., and Russ, J. (2004). Improving schools in socioeconomically disadvantaged areas—A review of research evidence. *School Effectiveness and School Improvement, 15,* 149-175.

Munro, D. J. (1969). *The concept of man in early China.* Stanford, CA: Stanford University Press.

Murnane, R. J. (1975). *The impact of school resources on the learning of inner city children.* Cambridge, MA: Ballinger.

Murnane, R. J., Willett, J. B., Bub, K. L., and McCartney, K. (2006). Understanding trends in the black-white achievement gaps during the first years of school. In G. Burtless and J. G. Rothenberg (Eds.), *Brookings-Wharton papers on urban affairs.* Washington, DC: Brookings Institution Press.

Murray, C. (2002). *IQ and income inequality in a sample of sibling pairs from advantaged family backgrounds.* Paper presented at the 114th annual meeting of the American Economic Association, Atlanta, GA.

Murray, C. (2003). *Human accomplishment: The pursuit of excellence in the arts and sciences, 800 B.C. to 1950.* New York: HarperCollins.

Murray, C. (2007a). Intelligence in the classroom. *Wall Street Journal.* Retrieved July 10, 2007, from http://www.opinionjournal.com/extra/?id=110009531.

Murray, C. (2007b, April). Jewish genius. *Commentary.* Retrieved October 17, 2007, from http://www.commentarymagazine.com/viewarticle.cfm?id=10855.

Myerson, J., Rank, M. R., Raines, F. Q., and Schnitzler, M. A. (1998). Race and general cognitive ability: The myth of diminishing returns to education. *Psychological Science, 9,* 139–142.

Myrdahl, G. (1944). *An American dilemma: The Negro problem and modern democracy.* New York: Harper.

Nakamura, H. (1964/1985). *Ways of thinking of eastern peoples: India, China, Tibet and Japan.* Honolulu: University of Hawaii Press.

Nakanishi, N. (1982). A report on the 'how do people spend their time survey' in 1980. *Studies of Broadcasting (An international annual of broadcasting science), 18,* 93–113.

National Aeronautics and Space Administration. (1978). *Anthropometric source book: Volume I: Anthropometry for Designers* (NASA Reference Publication 1024).

National Center for Education Statistics. (2000). *Pursuing excellence: Comparisons of international eighth-grade mathematics and science achievement from a U.S. perspective: 1995 and 1999.* Washington, DC: U.S. Department of Education.

National Endowment for the Arts. (2007). *To read or not to read: A question of national consequence.* Washington, DC: National Endowment for the Arts.

Neisser, U. (1996). Intelligence: Knowns and unknowns. *American Psychologist, 51,* 77–101.

Nettelbeck, T. (1998). Jensen's chronometric research: Neither simple nor sufficient but a good place to start. *Intelligence, 6,* 233–241.

Neuman, S. B., and Celano, D. (2001). Access to print in low-income and middle-income communities: An ecological study in four neighborhoods. *Reading Research Quarterly, 36,* 8–26.

Nisbett, R. E. (Ed.), (1992). *Rules for reasoning.* Hillsdale, NJ: Lawrence Erlbaum.

Nisbett, R. E. (2003). *The geography of thought: How Asians and Westerners think differently . . . and why.* New York: Free Press.

Nisbett, R. E., Fong, G. T., Lehman, D. R., and Cheng, P. W. (1987). Teaching reasoning. *Science, 238,* 625–631.

Norenzayan, A., Smith, E. E., Kim, B. J., and Nisbett, R. E. (2002). Cultural preferences for formal versus intuitive reasoning. *Cognitive Science, 26,* 653–684.

Nye, B., Jayne Zaharias, B. D., Fulton, C. M., Achilles, C. M., and Hooper,

R. (1994). *The lasting benefits study: A continuing analysis of the effect of small class size in kindergarten through third grade on student achievement test scores in subsequent grade levels* (Seventh grade technical report). Nashville: Center of Excellence for Research in Basic Skills, Tennessee State University.

Ogbu, J. U. (1978). *Minority education and caste: The American system in cross-cultural perspective.* New York: Academic Press.

Ogbu, J. U. (1991a). Immigrant and involuntary minorities in perspective. In M. Gibson and J. Ogbu (Eds.), *Minority status and schooling: A comparative study of immigrant and involuntary minorities.* New York: Garland.

Ogbu, J. U. (1991b). Low performance as an adaptation: The case of blacks in Stockton, California. In M. Gibson and J. Ogbu (Eds.), *Minority status and schooling: A comparative study of immigrant and involuntary minorities.* New York: Garland.

Ogbu, J. U. (1994). *Minority education and caste: The American system in cross-cultural perspective.* New York: Academic Press.

Ogbu, J. U. (2003). *Black American students in an affluent suburb: A study of academic disengagement.* Mahwah, NJ: Erlbaum Associates.

Oleson, P. J., Westerberg, H., and Klingberg, T. (2003). Increased prefrontal and parietal activity after training of working memory. *Nature Neuroscience, 7,* 75–79.

Organisation for Economic Co-operation and Development (OECD). (2000). *Knowledge and skills for life: First results from PISA 2000.* Paris: OECD.

Organisation for Economic Co-operation and Development (OECD).(2001). *Knowledge and skills for life: First results from the OECD Programme for International Student Assessment.* Paris: OECD.

Organisation for Economic Co-operation and Development (OECD). (2007). *Main science and technology indicators.* Retrieved August 20, 2007, from http://puck.sourceoecd.org/vl=4480226/cl=13/nw=1/rpsv/~3954/v207n1/s1/p1.

Ortar, G. (1967). Educational achievement of primary school graduates in Israel as related to their socio-cultural background. *Comparative Education, 4,* 23–35.

Osborne, J. W. (1997). Race and academic disidentification. *Journal of Educational Psychology, 89,* 728–735.

Otto, S. P. (2001). Intelligence: Historical and conceptual perspectives. In *International encyclopedia of the social and behavioral sciences.* Oxford: Perganon.

Oyserman, D., Bybee, D., and Terry, K. (2006). Possible selves and academic outcomes: How and when possible selves impel action. *Journal of Personality and Social Psychology, 91,* 188–204.

Pager, D. (2003). The mark of a criminal record. *American Journal of Sociology, 108,* 937–975.

Parra, E. J., Marcini, A., Akey, J., Martinson, J., et al. (1998). Estimating African American admixture proportion by use of population-specific alleles. *American Journal of Human Genetics, 63*, 1839–1851.

Parra, E. J., Kittles, R. A., and Shriver, M. D. (2004). Implications of correlations between skin color and genetic ancestry for biomedical research. *Nature Genetics, 36*, S54–S60.

Patai, R. (1977). *The Jewish mind.* New York: Scribners.

Patterson, O. (2006, March 26). A poverty of the mind. *New York Times.* Retrieved August 26, 2007, from http://select.nytimes.com/search/restricted/article?res=f30c1ef63c540c758eddaa0894de404482.

Pedersen, E., Faucher, T. A., and Eaton, W. W. (1978). A new perspective on the effects of first-grade teachers on children's subsequent adult status. *Harvard Educational Review, 48*, 1–31.

Peng, K. (1997). *Naive dialecticism and its effects on reasoning and judgment about contradiction.* Unpublished doctoral dissertation, University of Michigan, Ann Arbor.

Peters, M. (1995). Does brain size matter? A reply to Rushton and Ankney. *Canadian Journal of Experimental Psychology, 49*, 570–576.

Phillips, H., and Ebrahimi, H. (1993). Equation for success: Project SEED. In G. Cuevas and M. Driscoll (Eds.), *Reaching all students with mathematics.* Reston, VA: National Council of Teachers of Mathematics.

Phillips, M. (2000). Understanding ethnic differences in academic achievement: Empirical lessons from national data. In D. Grissmer and J. M. Ross (Eds.), *Analytic issues in the assessment of student achievement* (NCES 2000-050.) Washington, DC: U.S. Department of Education.

Phillips, M., Brooks-Gunn, J., Duncan, G., Klebanov, P. K., and Crane, J. (1998). Family background, parenting practices, and the black-white test score gap. In C. Jencks and M. Phillips (Eds.), *The black-white test score gap.* Washington, DC: Brookings Institution.

Pinker, S. (2002). *The blank slate: The modern denial of human nature.* New York: Viking.

Plomin, R., and Petrill, S. A. (1997). Genetics and intelligence: What's new? *Intelligence, 24*, 53–57.

Plomin, R., and Spinath, F. (2002). Genetics and general cognitive ability (g). *Trends in Cognitive Sciences, 6*, 169–176.

Pollitt, E., Gorman, K. S., Engle, P. L., Martorell, R., and Rivera, J. (1993). Early supplementary feeding and cognition. *Monographs of the Society for Research in Child Development, 58* (Serial No. 235).

Prabhakaran, V., Rypma, B., and Gabrieli, J. D. (2001). Neural substrates of mathematical reasoning: A functional magnetic resonance imaging study of neocortical activation during performance of the Necessary Arithmetic Operations Test. *Neuropsychology, 15*, 115–127.

Quindlen, A. (2008, May 27). The drive to excel. *New York Times.* Retrieved May 27, 2008, from http://query.nytimes.com/gst/fullpage.html?res=9B0DE0DA1638F931A15751C0A961948260&sec=&spon=&pagewanted=all.

Ramey, C. T., Campbell, F. A., Burchinal, M., Skinner, M. L., Gardner, D. M., and Ramey, S. L. (2000). Persistent effects of early childhood education on high-risk children and their mothers. *Applied Developmental Science, 4,* 2–14.

Ramey, S. L., and Ramey, C. T. (1999). Early experience and early intervention for children "at risk" for developmental delay and mental retardation. *Mental Retardation and Developmental Disabilities Research Reviews, 5,* 1–10.

Ramphal, C. (1962). *A study of three current problems in education.* India: University of Natal.

Raven, J. C., Court, J. H., and Raven, J. (1975). *Manual for Raven's Progressive Matrices and Vocabulary Scales.* London: Lewis.

Raz, N., Gunning, F. M., Head, D., Dupuis, J. H., McQuain, J., Briggs, S. D., et al. (1997). Selective aging of the human cerebral cortex observed in vivo: Differential vulnerability of the prefrontal gray matter. *Cerebral Cortex, 7,* 268–282.

Reeves, D. B. (2000). *Accountability in action: A blueprint for learning organizations.* Denver: Center for Performance Assessment.

Rockoff, R. (2004). The impact of individual teachers on student achievement: Evidence from panel data. *American Economic Review, 94,* 247–252.

Rosenholtz, S. J. (1985). Effective schools: Interpreting the evidence. *American Journal of Education, 93,* 352–388.

Ross, L. (1977). The intuitive psychologist and his shortcomings. In L. Berkowitz (Ed.), *Advances in experimental social psychology* (Vol. 10, pp. 173–220). New York: Academic Press.

Rothstein, R. (2004). *Class and schools: Using social, economic, and educational reform to close the black-white achievement gap.* Washington, DC: Economic Policy Institute.

Rouse, C., Brooks-Gunn, J., and McLanahan, S. (2005). Introducing the issue. *Future of Children, 15,* 5–13.

Rouse, C. E. (1998). Private school vouchers and educational achievement: An evaluation of the Milwaukee choice program. *Quarterly Journal of Economics, 113,* 553–602.

Rowe, D., Jacobsen, K., and Van den Oord, E. (1999). Genetic and environmental influences on vocabulary IQ: Parental education as a moderator. *Child Development, 70,* 1151–1162.

Rueda, M. R., Rothbart, M. K., McCandliss, B. D., Saccomanno, L., and Posner, M. I. (2005). Training, maturation, and genetic influences on the development of executive attention. *Proceedings of the National Academy of Sciences of the United States of America, 102,* 14931–14936.

Rushton, J. P. (1990). Race, brain size, and intelligence: A rejoinder to Cain and Vanderwolf. *Personality and Individual Differences, 11,* 785–794.

Rushton, J. P., and Jensen, A. R. (2005). Thirty years of research on race differences in cognitive ability. *Psychology, Public Policy and Law, 11,* 235–294.

Rushton, J. P., and Jensen, A. R. (2006). The totality of available evidence shows the race IQ gap still remains. *Psychological Science, 17*, 921–922.

Rutter, J. M. (2000). Comments in discussion on James R. Flynn. In G. R. Bock, J. Goode, and K. Webb (Eds.), *The nature of intelligence.* Novartis Foundation Symposium 233. New York: Wiley.

Sampson, R. J., Morenoff, J. D., and Raudenbush, S. (2005). Social anatomy of racial and ethnic disparities in violence. *American Journal of Public Health, 95.*

Sanders, W. L., and Horn, S. P. (1996). Research findings from the Tennessee Value-Added Assessment Model (TVAAM) database: Implications for educational evaluation and research. *Journal of Personnel Evaluation in Education, 12*, 247–256.

Sarason, S. B. (1973). Jewishness, blackishness, and the nature-nurture controversy. *American Psychologist, 28*, 963–964.

Sarton, G. (1975). *Introduction to the history of science.* Huntington, NY: R. E. Krieger.

Scarr, S. (1981). *Race, social class, and individual differences in IQ: New studies of old issues.* Hillsdale, NJ: Lawrence Erlbaum.

Scarr, S. (1992). Developmental theories for the 1990s: Development and individual differences. *Child Development, 63*, 1–19.

Scarr, S., and McCartney, K. (1983). How people make their own environments: A theory of genotype→environment effects. *Child Development, 54*, 424–435.

Scarr, S., Pakstis, A. J., Katz, S. H., and Barker, W. B. (1977). Absence of a relationship between degree of white ancestry and intellectual skills within a black population. *Human Genetics, 39*, 69–86.

Scarr, S., and Weinberg, R. A. (1976). IQ test performance of black children adopted by white families. *American Psychologist, 31*, 726–739.

Scarr, S., and Weinberg, R. A. (1983). The Minnesota adoption studies: Genetic differences and malleability. *Child Development, 54*, 260–267.

Scarr-Salapatek, S. (1971). Race, social class, and IQ. *Science, 174*, 1285–1295.

Schiff, M., Duyme, M., Stewart, J., Tomkiewicz, S., and Feingold, J. (1978). Intellectual status of working-class children adopted early in upper-middle class families. *Science, 200*, 1503–1504.

Schneider, D. (2006). Smart as we can get? *American Scientist, 94*, 311–312.

Schoenemann, P. T., Budinger, T. F., Sarich, V. M., and Wang, W. S.-Y. (1999). Brain size does not predict general cognitive ability within families. *Proceedings of the National Academy of Science, 97*, 4932–4937.

Schoenthaler, S. J., Amos, S. P., Eysenck, H. J., Peritz, E., and Yudkin, J. (1991). Controlled trial of vitamin-mineral supplementation: Effects on intelligence and performance. *Personality and Individual Differences, 12*, 351–362.

Schweinhart, L. J., Montie, J., Xiang, Z., Barnett, W. S., Belfield, C. R., and

Nores, M. (2005). *Lifetime effects: The High/Scope Perry Preschool Study through age 40*. Ypsilanti, MI: High/Scope Foundation.

Schweinhart, L. J., and Weikart, D. P. (1980). *Young children grow up: The effects of the Perry Preschool Program on youths through age 15* (No. 7). Ypsilanti, MI: High Scope Press.

Schweinhart, L. J., and Weikart, D. P. (1993, November). Success by empowerment: The High/Scope Perry Preschool Study through age 27. *Young Children, 48,* 54–58.

Schwidetsky, I. (1977). Postpleistocene evolution of the brain. *American Journal of Physical Anthropology, 45,* 605–611.

Sherman, M., and Key, C. B. (1932). The intelligence of isolated mountain children. *Child Development, 3,* 279–290.

Shuey, A. M. (1966). *The testing of Negro intelligence* (2nd ed.). New York: Social Science Press.

Skuy, M., Gewer, A., Osrin, Y., Khunou, D., Fridjhon, P., and Rushton, J. P. (2002). Effects of mediated learning experience on Raven's matrices scores of African and non-African university students in South Africa. *Intelligence, 30,* 221–232.

Slavin, R. E. (1995). *Cooperative learning: Theory, research and practice* (2nd ed.). Boston: Allyn and Bacon.

Slavin, R. E. (2005). *Show me the evidence: Effective programs for elementary and secondary schools*. Baltimore, MD: Johns Hopkins University, Center for Data-Driven Reform in Education.

Snyderman, M., and Rothman, S. (1988). *The IQ controversy, the media and public policy*. New Brunswick, NJ: Transaction Books.

Sobel, M. (1987). *The world they made together: Black and white values in eighteenth-century Virginia*. Princeton, NJ: Princeton University Press.

Sonne-Holm, S., Sorensen, T. I., Jensen, G., and Schnohr, P. (1989). Influence of fatness, intelligence, education and sociodemographic factors on response rate in a health survey. *Journal of Epidemiology and Community Health, 43,* 369–374.

Sowell, T. (1978). Three black histories. In T. Sowell (Ed.), *Essays and data on American ethnic groups*. New York: Urban Institute.

Sowell, T. (1981). *Ethnic America: A history*. New York: Basic Books.

Sowell, T. (1994). *Race and culture: A world view*. New York: Basic Books.

Sowell, T. (2005). *Black rednecks and white liberals*. San Francisco: Encounter Books.

Steele, C. M. (1997). A threat in the air: How stereotypes shape intellectual identity and performance. *American Psychologist, 52,* 613–629.

Steele, C. M., and Aronson, J. (1995). Stereotype threat and the intellectual test performance of African Americans. *Journal of Personality and Social Psychology, 69,* 797–811.

Steele, C. M., Spencer, S., and Aronson, J. (2002). Contending with group image: The psychology of stereotype and social identity threat. In

M. Zanna (Ed.), *Advances in Experimental Social Psychology, Vol. 37.* New York: Academic Press.

Sternberg, R. J. (1999). The theory of successful intelligence. *Review of General Psychology, 3,* 292–316.

Sternberg, R. J. (2006). The Rainbow Project: Enhancing the SAT through assessments of analytic, practical, and creative skills. *Intelligence, 34,* 321–350.

Sternberg, R. J. (2007a, July 6). Finding students who are wise, practical, and creative. *Chronicle of Higher Education.* Retrieved October 19, 2007, from http://chronicle.com/subscribe/login?url=/weekly/v53/i44/44b01101.htm.

Sternberg, R. J. (2007b). Intelligence and culture. In S. Kitayama and D. Cohen (Eds.), *Handbook of cultural psychology.* New York: Guilford Press.

Sternberg, R. J., Wagner, R. K., Williams, W. M., and Horvath, J. A. (1995). Testing common sense. *American Psychologist, 50,* 912–927.

Stevenson, H. W., Lee, S. Y., Chen, C., Stigler, J. W., Hsu, C. C., and Kitamura, S. (1990). Contexts of achievement: A study of American, Chinese and Japanese children. *Monographs for the Society for Research in Child Development, 55* (1-2, Serial No. 221).

Stevenson, H. W., and Stigler, J. W. (1992). *The learning gap: Why our schools are failing and what can we learn from Japanese and Chinese education.* New York: Summit Books.

Stoolmiller, M. (1999). Implications of the restricted range of family environments for estimates of heritability and nonshared environment in behavior-genetic adoption studies. *Psychological Bulletin, 125,* 392–409.

Streissguth, A. P., Barr, H. M., Sampson, P. D., Darby, B. L., and Martin, D. C. (1989). IQ at age 4 in relation to maternal alcohol use and smoking during pregancy. *Developmental Psychology, 25,* 3–11.

Sugar, B. R. (2006, February 19). Punching through. *New York Review of Books,* 19.

Tang, Y., Ma, Y., Wang, J., Fan, Y., et al. (2007). Short-term meditation training improves attention and self-regulation. *Proceedings of the National Academy of Sciences of the United States of America, 104,* 17152–17156.

Taylor, H. F. (1980). *The IQ game: A methodological inquiry into the heredity-environment controversy.* New Brunswick, NJ: Rutgers University Press.

Thernstrom, S., and Thernstrom, A. (1997). *America in black and white: One nation indivisible.* New York: Simon and Schuster.

Tizard, B., Cooperman, A., and Tizard, J. (1972). Environmental effects on language development: A study of young children in long-stay residential nurseries. *Child Development, 43,* 342–343.

Tough, P. (2007, June 10). The class-consciousness raiser. *New York Times Magazine,* 52.

Turkheimer, E., Haley, A., Waldron, M., D'Onofrio, B., and Gottesman, I. I. (2003). Socioeconomic status modifies heritability of IQ in young children. *Psychological Science, 14,* 623–628.

U.S. Census Bureau. (2006). Retrieved December 4, 2006, from http://www .census.gov/population/www/socdemo/education/cps2006.html.

U.S. Department of Education. (1998). *TIMSS [Third International Mathematics and Science Study] 12th-grade report: Questions and answers.* Washington, DC: U.S. Department of Education.

U.S. Department of Education. (2008). What Works Clearinghouse. Retrieved May 25, 2008, from http://ies.ed.gov/ncee/wwc/.

U.S. Department of Health and Human Services. (2005). *Head Start impact study: First year findings.* Washington, DC: Administration for Children and Families.

U.S. Department of Health and Human Services. (2006). National Immunization Survey. Washington, DC: U.S. Department of Health and Human Services.

U.S. Office of Personnel Management. (2006). Retrieved December 8, 2006, from http://www.opm.gov/feddata/demograp/demograp.asp.

van IJzendoorn, M. H., Juffer, F., and Klein Poelhuis, C. W. (2005). Adoption and cognitive development: A meta-analytic comparison of adopted and nonadopted children's IQ and school performance. *Psychological Bulletin, 131,* 301–316.

Van Loon, A. J. M., Tijhuis, M., Picavet, H. S. J., Surtees, P. G., and Ormel, J. (2003). Survey non-response in the Netherlands: Effects on prevalence estimates and associations. *Annals of Epidemiology, 13,* 105–110.

van Zeigl, J., Mesman, J., van IJzendoorn, M. H., Bakersman-Kranenburg, M. J., and Juffer, F. (2006). Attachment-based intervention for enhancing sensitive discipline in mothers of 1–3-year-old children at risk for externalizing behavior problems: A randomized controlled trial. *Journal of Consulting and Clinical Psychology, 74,* 994–1005.

Verhulst, F. C., Althaus, M., and Versluis-den Bieman, H. J. M. (1990). Problem behavior in international adoptees: I. An epidemiological study. *Journal of American Academy of Child and Adolescent Psychiatry, 29,* 518–524.

Vernon, P. E. (1982). *The abilities and achievements of Orientals in North America.* New York: Academic Press.

Walton, G. M., and Cohen, G. L. (2007). A question of belonging: Race, social fit, and achievement. *Journal of Personality and Social Psychology, 92,* 82–96.

Wasik, B. H., Ramey, C. T., Bryant, D. M., and Sparling, J. J. (1990). A longitudinal study of two early intervention strategies: Project CARE. *Child Development, 61,* 1682–1696.

Watanabe, M. (1998). *Styles of reasoning in Japan and the United States: Logic of education in two cultures.* Paper presented at the American Sociological Association, San Francisco, CA.

Waters, M. C. (1999). *Black identities: West Indian immigrant dreams and American realities*. Cambridge, MA: Harvard University Press.

Webster, W. J., and Chadbourn, R. A. (1992). *The evaluation of Project SEED*. Dallas: Dallas Independent School District.

Weinberg, R. A., Scarr, S., and Waldman, I. D. (1992). The Minnesota Transracial Adoption Study: A follow-up of IQ test performance at adolescence. *Intelligence, 16*, 117–135.

Wicherts, J. M., Dolan, C. V., Carlson, J. S., and van der Maas, H. L. J. (2008). *IQ test performance of Africans: Mean level, psychometric properties, and the Flynn effect*. Unpublished manuscript, Amsterdam: University of Amsterdam.

Willerman, L., Naylore, A. F., and Myrianthopoulos, N. C. (1974). Intellectual development of children from interracial matings: Performance in infancy and at 4 years. *Behavior Genetics, 4*, 84–88.

Williams, W. M. (1998). Are we raising smarter children today? School- and home-related influences on IQ. In U. Neisser (Ed.), *The rising curve: Long-term changes in IQ and related measures*. Washington, DC: American Psychological Association.

Witty, P. A., and Jenkins, M. D. (1934). The educational achievement of a group of gifted Negro children. *Journal of Educational Psychology, 25*, 585–597.

Witty, P. A., and Jenkins, M. D. (1936). Inter-race testing and Negro intelligence. *Journal of Psychology, 1*, 188–191.

Woods, R. P., Freimer, N. B., De Young, J. A., Fears, S. C., et al. (2006). Normal variants of Microcephalin and ASPM do not account for brain size variability. *Human Molecular Genetics, 15*, 2025–2029.

Zweig, S. (1943/1987). *The world of yesterday*. London: Cassell.

CREDITS

Figure 1.1   Reprinted with permission from Flynn, 2007, p. 16. Copyright James R. Flynn.

Figure 1.2   Reprinted with permission from Cattell, 1987, p. 206. Copyright Elsevier Science Publishers.

Box 1.1   Reprinted with permission from Flynn, 2007, p. 5. Copyright James R. Flynn.

Table 2.1   Reprinted with permission from Devlin, Daniels, and Roeder, 1997, p. 469. Copyright Nature Publishing Group.

Figure 3.1   Reprinted with permission from Flynn, 2007, p. 8. Copyright James R. Flynn.

Figure 3.2   Reprinted with permission from Eicholz, R., 1991, p. 56. Copyright Pearson Education.

Figure 7.1   Reprinted with permission from Knudsen, Heckman, Cameron, and Shonkoff, 2006, p. 10156. Copyright National Academy of Sciences of the United States of America.

Figure 8.1   Reprinted with permission from Masuda and Nisbett, 2001, p. 924. Copyright American Psychological Association.

INDEX

Page numbers in *italics* refer to boxes and figures. Page numbers beginning with 237 refer to notes.